Philosophical Urbanism

Abraham Akkerman

Philosophical Urbanism

Lineages in Mind-Environment Patterns

palgrave
macmillan

Abraham Akkerman
Department of Geography and Planning
University of Saskatchewan
Saskatoon, SK, Canada

Department of Philosophy
University of Saskatchewan
Saskatoon, SK, Canada

ISBN 978-3-030-29084-9 ISBN 978-3-030-29085-6 (eBook)
https://doi.org/10.1007/978-3-030-29085-6

This Palgrave Macmillan imprint is published by the registered company Springer Nature Switzerland AG.
The registered company address is: Gewerbestrasse 11, 6330 Cham, Switzerland

To Amitai and Azariah

Contents

Abbreviations

BP	Before Present
IACE	Iron Age Cold Epoch, c. 900–c. 300 BCE (North Atlantic region)
Kya	Thousand years ago
LGP	Last Glacial Period, c. 115,000–c. 12,000 BP (largely, northern hemisphere)
LIA	Little Ice Age, c. 1350–c. 1850 (largely, Eurasia)
Mya	Million years ago
NDT	Neolithic Demographic Transition, c. 12,000 BP till the onset of the Bronze Age

Periodicity

Climate history periods

Pleistocene, or Ice Age: 2.5 mya–12,000 BP
LGP, or Last Glacial Period, c. 115,000–12,000 BP
Holocene, the current climatic and geological era, starting 12,000 BP
Last Glacial Maximum: 31,000–16,000 BP

Eurasia human prehistory periods

Upper Paleolithic, c. 50,000–12,000 BP
Late Upper Paleolithic 18,000–12,000 BP
Gravettian, c. 30,000–20,000 BP
Neolithic, c. 12,000–c. 6000 BP
Copper Age (Chalcolithic), c. 6000–c. 4500 BP
Bronze Age, c. 2500–c. 1200 BCE

Mediterranean/western European history periods

Iron Age, c. 1200–c. 700 BCE
Classical Antiquity, c. 700–c. 500 CE
  Hellenic Antiquity, c. 300–c. 100 BCE
  Late Antiquity/Early Middle Ages, c. 250–c. 600 CE

High Middle Ages/Romanesque, c. 900–c. 1250
Late Middle Ages, c. 1250–c. 1350
Renaissance/Early modernity, c. 1350–c. 1600
Enlightenment/Scientific Revolution, c. 1600–c. 1800
Modernity, c. 1800–c. 1950
Late modernity, c. 1950–

List of Figures

List of Tables

CHAPTER 1

Philosophical Urbanism of Walter Benjamin

Introduction and Summary

It was Herbert Spencer who in the second half of the nineteenth century introduced the idea that the environment and organisms within it are in an ongoing evolutionary feedback (Spencer 1864: 82–93; Pearce 2010). Less than a century later, Walter Benjamin, most likely unaware of Spencer's work, had addressed the seemingly narrower scope of feedback of minds interacting with their *built* environment in the rise of modernity. The context of Benjamin's notion of interaction between minds and the built environment ought to be seen as encompassing also climate, or weather, and demography, or the crowd. Thus modernity, as an intellectual achievement epitomized by the scientific revolution and the rise of the public sphere, is perceived here also within the twofold framework of climate and demography. The climatic event spanning much of modernity is known as the Little Ice Age while, on the other hand, late modernity had been marked by the demographic transition event of the gradual and uneven drop in fertility and mortality in human populations. Both climate and demography have been intertwined within the urban environment. Late modernity has been marked by climate change induced, at least partially, by 'heat islands' of intense urbanization since the nineteenth century, caused by human populations. Gradually urbanizing throughout history, humans during late modernity have escalated their urban habitat, expanding it into sprawling metropolitan areas throughout much of the world today.

A. Akkerman, *Philosophical Urbanism*,
https://doi.org/10.1007/978-3-030-29085-6_1

The demographic context of modernity has been mainly that of the urban crowd, a human multitude continually redefining the use of urban space, particularly urban niches seen as public spaces. The urban crowd, and the associated notion of the *stranger*, signify culmination in a demographic process that has been ingrained within the unfolding of civilization, as well as within the history of climate. Civilization is the latest stage in the evolution of anatomically modern humans who emerged between 300,000 and 200,000 BP. The history and prehistory of civilization can be perceived as set between two demographic transitions, about 12,000 years apart: one, the Neolithic Demographic Transition (NDT) defined by transition to sedentarism, agriculture and domestication of animals, accompanied by higher fertility (Bocquet-Appel 2008), and the other, the modern demographic transition consisting in the unequal transformation of fertility and mortality, from high to low (Caldwell 2006: 157–216).

The NDT, also referred to as the Neolithic Agricultural Revolution, began about 12,000 BP, at the onset of the Neolithic, the last stage of the Stone Age which started about 3.4 million BP. The Neolithic was sparked by the Holocene, the current climatic and geological era that started c. 12,000 BP, marking the end of the Last Glacial Period on Earth, c. 115,000 BP to 12,000 BP. The coterminous onset of the Neolithic, the NDT and the Holocene, is no coincidence. As climate warmed, human nomadic hunters of game and gatherers of fruits and vegetables inaugurated domestication of animals and land cultivation for the domestication of plants. The human production of food crops and domestication of animals had led to improved nutrition and higher fertility and came to play a key role in the history of civilization. The storing of grains, and the particular use of donkeys and horses for work in agriculture and construction, was behind the rise of first urbanized civilizations during the Bronze Age: the Harappan of the Indus Valley, fourth to second century BCE, and Sumer of southern Mesopotamia, fourth to third century BCE.

A considerable role in the unfolding of interaction between anatomically modern humans and the environment played gender awareness. Discernment of gender and of human aging is the founding tenet at the base of demography. Gender and aging are also the manifest features of constancy and change, respectively, that had left their imprints on the perception of humans in the midst of their kin. The awareness of two mutually exclusive gender types, the male and the female, had led the archaic men and women to an archetypal gender myth, through which they explained to themselves the founding and the working of the universe.

Such acumen had led, in turn, to the projection of gender traits upon human interpretation of the environment, and upon the founding of human-built environments.

The projection of gender upon built environments has been through two foundational features, void and volume. Voids had defined primordial public or communal spaces within the early built environment: the round enclosure of fertility rites and renewal ceremonies of the Neolithic, that is, designed voids constituting largely feminine projection upon the environment. The masculine projection had been manifest during the Neolithic and the later Bronze Age, initially in the placement and configuration of standing stones, and later in the construction of edifices, particularly then defensive features of human settlements such as walls and ramparts, along with forts and citadels. Literary documentation of the archetypal projection of gender upon the environment had been the narrative of the Garden and the Citadel, while during modernity an ardent account of environmental gender projection came in 'The Castle in the Forest,' a chapter of Carl Jung's *Red Book* (1913/2009).

1.1 Modernity: Walter Benjamin on Mind and the Urban Environment

Modernity has been usually viewed as an alignment of social and cultural norms that had gradually evolved through the European Renaissance and the Enlightenment. Coming to a close in the second half of the twentieth century, modernity's dawn has been variously assigned across periods ranging from the fifteenth to the eighteenth century. The variance in the definition has been bound to differing assignment of significance to events that had led to scientific and technological progress, transition to capitalism or advance toward new social arrangements, intensifying urbanization and the nation-state. While such specifics constitute less than a unanimous agreement, there is little dispute that processes that had marked the greater thrust of modernity, were profoundly accentuated by the rationalism of René Descartes in Holland, France and Germany; the astronomy of John Kepler in Austria and Bohemia and the physics of Galileo Galilei in the north of Italy, all within the first decades of the seventeenth century.

Modernity's sequel that came on the heels of these developments was largely confined to the two inhabited continental masses delineating the northern Atlantic, mainly Europe, north of the Alps and west of the

Carpathian Mountains and the eastern seaboard of North America. The case to argue is that the philosophy and thought of modernity that have emerged from within this geographic region, have been significantly influenced by its environment, while in turn, they shaped—for better or worse—the human impact cast on the very same environment as well as beyond it.

The stance of a dynamic and intrinsically indeterminist mind-environment feedback may have been stated first by the physician Julien Offroy de La Mettrie in his *L'Homme Machine* (1748/1994):

> [W]ho knows if the reason for man's existence might not lie in this very existence itself? Perhaps he was cast by chance into a tiny corner of the earth, knowing neither how or why [...] (La Mettrie 1748/1994: 54)

La Mettrie's embrace of indeterminism and chance in situating humans in the world had come as a contrarian viewpoint to environmental determinism, a prevalent belief during the Enlightenment, associating mentalities with climate or with geographic latitude. Environmental determinism gained momentum in early modernity thanks to Charles de Montesquieu who, in *The Spirit of the Laws* (1749/1949), put forward statements that became mainstream standard in the mid-eighteenth century. Associating climate with behavior and mindset Montesquieu claimed to explain some legislation of the West Indies as "arising from the laziness of the climate" (Montesquieu 1749/1949: 226). With such disposition Montesquieu set the ground to further similar pernicious arguments, universally considered today as not only questionable but also deplorable. It is due to the implicit bigotry as well as a methodically simplistic view claiming causal impact of climate and geography upon the makeup of people's character, that by the end of the twentieth-century environmental determinism has been largely rejected, in the discipline of geography in particular (Peet 1985). At the turn of the twenty-first century, a revival of sorts of environmental determinism steered clear of earlier claims pursuing impact of the environment upon human cognition and focused on the force of environmental factors upon cultures and civilizations (Diamond 1999: 53–66).

The present study is an attempt to redress the controversial notion of relation between mind and the environment, specifically focusing upon the *built* environment, and pointing to a context that is entirely indeterministic. Mind-city interaction has been pursued by philosophers and geographers sporadically since the turn of the twentieth century, but by the century's end, it was given only a scant consideration. This has been so in spite of urbanization having been perhaps the most prominent *environmental* issue from which deliberations about modernity have emanated. Urbanization has been seen as a driving force behind nuances of modernity while also other, less acknowledged, environmental agents had been seen crucial to interaction with cognition and reasoning, by means of urbanization, during this historical period. Two such other environmental issues had been, as Michel Foucault had alluded half a century ago, demography and climate (Foucault 1978/2009: 12–29). Along with urbanization itself, weather and demography carried marked weight upon modernity, but cumulatively also upon modernity's modes of contemplation. As intertwined processes setting the environmental context within which modern thought has evolved, demography, climate and urbanization, however, had received only occasional attention (Behringer 2010: 85–120; Johnson-Hanks 2008).

Mind-environment mutual interaction, or the more recent variety, mind-city interaction, seems to have been pronounced and detectable in the North Atlantic landmass. The survey of cases showing such interaction during modernity, as well as inquiry into such reciprocity for all history and geography as framing the course of civilization, has been labeled *philosophical urbanism* (Akkerman 2013). As opposed to environmental or architectural determinism, claiming a one-directional causal impact from the environment upon mind and behavior, philosophical urbanism is an entirely non-deterministic outlook pointing to effervescent interaction between mind and the environment, ongoing over historic time and geographic space.

The notion of modern mind in *interaction* with environmental patterns had become the subject of scrutiny almost a century ago, ushered by the cultural critic Walter Benjamin. In his *Arcades Project*, Benjamin pointed to the mutual and ongoing feedback between mind and city-form as the most prominent aspect of material culture (Buck-Morrs 1990: 114). Benjamin suggests that civilization creates and re-creates its environments through ideal images of earlier ages. Such proposition could be exemplified in several instances. Thus, for example, pointing to the transition between the Middle Ages and the Renaissance, Erwin Panofsky had shown that scholastic thought was a spatio-cerebral process of transaction between architecture of the Gothic and the monastic mind, setting the ground for systematic reasoning and the advent of the Enlightenment (Panofsky 1957). This study attempts to open a discussion that will address the unfolding, since prehistory, of the feedback progression between mind and the built environment, as ultimately forging modernity and postmodernity.

Benjamin's is a dynamic stance about the relation between mind and city-form, an entirely non-deterministic, and in particular, non-Marxian attitude. The frequent labeling of Benjamin as Marxian or Marxist (Leslie 2007; Lunn 1982: 215–241) stems from his series of observational essays or notes that had commenced in the *Arcades Project*, and that were occasionally tainted with a Marxist outlook. It is also true that Benjamin's later dissociation from Marxism, documented in his 1940 pamphlet, *On the Philosophy of History* (Ferris 2008: 130–135) had never been entirely absorbed into his urbanist outlook. Benjamin predicates, however, a haphazard process of mutual, environ/*mental* reciprocation from which he sees twentieth-century city-form emerging. To a Marxist such proposition could be nothing short of heresy.

Benjamin's urbanist thought is largely a reflection on the urban poetry of Charles Baudelaire about the people and their built environment of nineteenth century's Paris (Benjamin 1935/1983). Newly carved boulevards replacing old, crooked streets were one focus of Baudelaire, the other, capturing also Benjamin's observations decades later, were Parisian shopping arcades with glazed rooftops which made their debut in the French capital during the early years of the nineteenth century, partly in response to the atrocious disrepair of its streets.

Similar to their Parisian counterparts, London's streets too had been enduring neglect since the Middle Ages, and decrepit streetscapes had been a common sight in London during much of the nineteenth century. Following the Great Fire of 1666 several plans for the rebuilding of the city were put forward, none of them pursued, however, due to political turmoil and property ownership claims. In 1806 John Nash, a landscape architect, was appointed to become the King's Commissioner of Woods and Forests and was charged with the design of an arterial road east of Mayfair, an affluent and fashionable section of London. The newly designed thoroughfare would separate Mayfair from ramshackle, poor neighborhoods further east. With the strong support of the Prince Regent, later King George IV, in 1813 the parliament had approved Nash's design for what was to be known as New Street, conditioned by demolishing much of the seventeenth century's Swallow Street and numerous adjacent old lanes, remnants of London's history. By 1819 the designed thoroughfare, almost 1 kilometer in length, was established as a fashionable boulevard, the world's first retail shopping street (Fig. 1.1). Serving as the

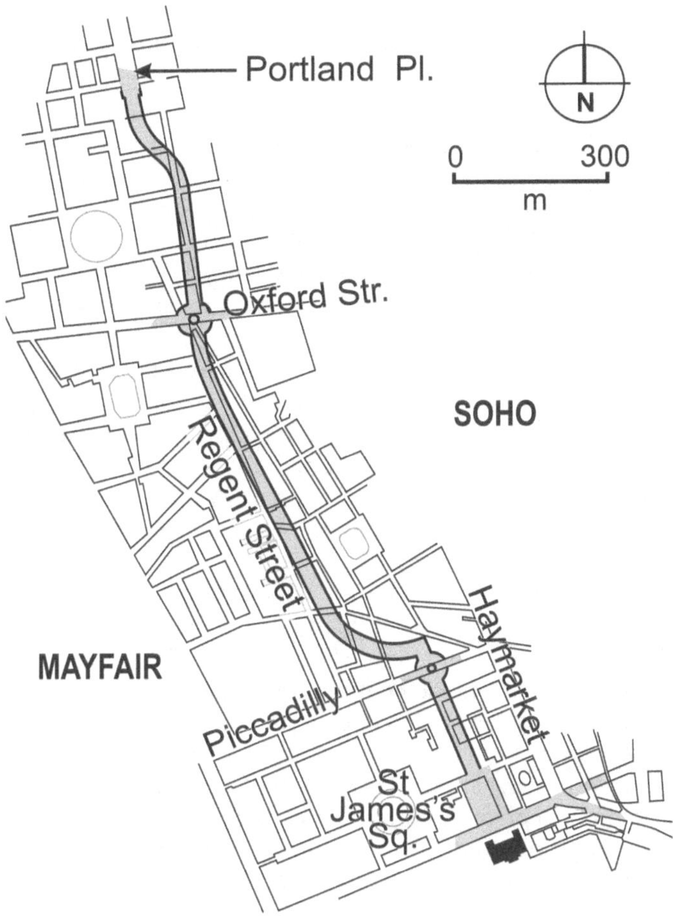

Fig. 1.1 Plan of Regent Street and adjacent areas, 1819. (Drawn by Paul Van Pul)

dividing line between Soho, the socially and economically deprived neighborhood further to the east and south, and the fashionable squares and streets of Mayfair to the west, it was soon renamed Regent Street (Morris 2013: 271–272). Sporting a colonnade made of cast-iron columns, offering protection from bad weather, Regent Street had attracted scores of pedestrians, thus becoming also a home to numerous cafés.

Regent Street was the venue in a short story by Edgar Allan Poe, 'The Man of the Crowd' (1840/2004), about a café onlooker who follows a stranger singled out on this fashionable and busy London street. And it was chiefly this story, along with the newly emerging urban environment of Paris, which had inspired Charles Baudelaire to write his notorious book of poems, *The Flowers of Evil*. Following its publication in 1857 some of the poems, and eventually the entire book, were banned by authorities, as offense against public morals and religion. In the second edition (1861) a section of 18 poems was added, most of them written during the commencement of the mid-nineteenth century's renovation of Paris. The section, entitled *Parisian Scenes*, reflects on the diurnal cycle of Parisian streetscapes from within which Baudelaire's feelings of anonymity and alienation in the newly modernized city emerge. As if following his own premise in the *Arcades Project* on the reciprocation between thought and the built environment, Benjamin had absorbed both Baudelaire and Poe, within the urban milieu of Paris and Berlin in the third and fourth decade of the twentieth century, reflecting back on Baudelaire's urban observation almost a century earlier.

In 1924 Winston Churchill, as the Chancellor of the Exchequer, had remarked in a speech to the Architectural Association, "we make buildings, and afterwards the buildings make us" (Duffy 1974). As a profoundly scholarly notion, the idea of a feedback progression that mutates minds as well as their built environments through time had emerged with the launching of cultural geography by Carl O. Sauer, a decade before Benjamin's work. In 'The Morphology of Landscape,' Sauer (1925) advanced the concept of cultural landscape as the imposition of culture upon nature. Culture, as defined by the shared myths, beliefs and behavioral standards, manifests itself in a human intervention upon *natural* landscape. Culture and one of its more important aspects, technology, have been changing the lived environment of humans, primarily through considerations of expediency related to the human body, while bodily experiences have been changing accordingly and, in turn, have often driven advances of further technological change. Landscapes modified by human action, as an aspect of culture, thus impact culture itself, whereby the feedback interaction between culture and landscape, as a progression in time, defines much of the history of civilization.

Quite independent of Benjamin's earlier work Sauer's cultural geography had set in motion a century-long parallel discourse. The geographic inquiry by David Lowenthal into psychoanalytic aspects of landscape had pointed to environmental judgment and action as tied to

> perceptions forged by preconception and experience, colored by taste and preference, variegated by culture and personality, amplified by memory and diminished by amnesia. (Lowenthal 1994)

To Lowenthal, time and the temporal aspect of a landscape, as often manifest in nostalgia arising from the memory of the past, constitute the ingredient that gives meaning to the presence of humans within their physical environments. In a 1975 paper Lowenthal, inadvertently echoing Benjamin, specifically addressed the shaping of physical landscapes through the myths and beliefs to which these very landscapes gave rise in the first place:

> As we erode and alter the inherited past, we more and more contrive our own. Creatures of historical processes beyond our control, we shape landscapes and artifacts to conform with illusory histories, public and private, that gratify our tastes. (Lowenthal 1975)

Focusing on city-form specifically, Benjamin discerns the two added environmental concerns of weather and population. To Benjamin weather is one of the highest "most genial" manifestations of cosmic forces, taken for granted by the common man, yet profoundly affecting his mind:

> Nothing is more characteristic than that precisely this most intimate and mysterious affair, the working of the weather on humans should have become the theme of their emptiest chatter. [...] How fine the ironic overcoming of this attitude in the story of the splenetic Englishman who wakes up one morning and shoots himself because it is raining. (Benjamin 1933/1999: 101–102)

1.2 The Environ/*mental* Context of Benjamin's *Arcades Project*: Echoes of the Little Ice Age

Notwithstanding his faith in the "cosmic" power weather wields upon mind, the foundational premise implicit in the work of Benjamin is that it is mainly the built environment that constitutes continued and sustained source of impact upon cognition. The return impact upon the built environment is not only through deliberate design, but mainly through the conglomerate of human bodies and minds subjected to such recurrent change. Increasingly between the Renaissance and late modernity, the intermediary in the interaction between the individual and city-form, has been the urban crowd.

Benjamin beholds an intangible power of the urban crowd. A strong reinforcement to his own position Benjamin finds in the correspondence of Charles Dickens in the preceding century:

> I cannot express how much I want [the streets ...] It seems as if they supplied something to my brain which it cannot bear, when busy, to lose [...] My figures seem disposed to stagnate without crowds about them. (Dickens quoted in Benjamin 1940/1999: 426)

Dickens' statement reflects on the urban crowd as something far more powerful than sheer demographic count. But the urban streetscape, too, even without the people within it, can be beyond mere analysis of a city plan or a realty consideration. Henri Bergson in his *Introduction to Metaphysics* evokes the experience of a person walking through streetscape as contrasted with viewing the same streetscape in a series of snapshot images:

> Were all the photographs of a town, taken from all possible points of view, to go on indefinitely completing one another, they would never be equivalent to the solid town in which we walk about. Were all the translations of a poem into all possible languages to add together their various shades of meaning and, correcting each other by a kind of mutual retouching, to give a more faithful image of the poem they translate, they would yet never succeeded in rendering the inner meaning of the original. (Bergson 1903/1912: 5)

Varying degree of environmental cognitive effect of streetscapes may be assumed for all individuals, but taken jointly, the two observations by Dickens and Bergson, provide some insight into the particular fascination of modern city-form by the likes of Edgar Allan Poe, Charles Baudelaire and Walter Benjamin.

Similarly too, the reaction of mood to weather has been the subject of folk belief, possibly since at least the Middle Ages. Recognition of the climatic context of city-form, whether analytically focused or broadly studious, is thus of added significance. This is so also because marked increases of precipitation through all seasons were the characteristics of much of modernity, throughout the northern Atlantic especially. The Little Ice Age (LIA), variously demarcated as spanning a period between 300 and 500 years, ending at mid- to late-nineteenth century, was a cooling event characterized by drop of average temperature, intermittently also exacerbated by lengthy winters, and extreme chill. It was during the closure of the LIA that also the concept of the urban crowd, as a disparaging notion, had been introduced into philosophy through the works of two foremost existentialist thinkers of the nineteenth century: Søren Kierkegaard's *The Crowd is Untruth* (1859) and Friedrich Nietzsche's *Beyond Good and Evil* (1886/1997). This was a period when many European cities had continued to experience rapid increase in population due to incessant migration from countryside. This demographic event started in the sixteenth and seventeenth centuries and, tempered by improvement in public health along with attendant decline in mortality, it yielded the modern demographic transition of the eighteenth and nineteenth centuries in Europe.

In contrast, at the close of the Middle Ages European urbanization was far from fostering population increase. Throughout the Middle Ages, European cities arose as interchanges of transport and commerce, as political and military strongholds or as religious and administrative hubs. At the turn of the fourteenth-century populations of the largest cities in Europe—Cologne, Prague, Florence, London or Paris—had ranged from 40,000 to 275,000, but densely populated, walled cities had increasingly carried growing risk of contagion. Rome, which had almost 1 million inhabitants at the beginning of the Common Era, due to internal Church rivalries that culminated in papal schism of the fourteenth century, and the ensuing neglect, had seen its population drop to less than 20,000 during this time (Logan 2002: 275–286, 297–307). Other cities too, experiencing violence in armed conflicts, as well as famine and epidemics, saw their populations decline rapidly during the fourteenth century. The onset of the Little Ice Age had further exacerbated hardship and loss of life, mainly due to food shortages, while population decline had contributed to reforestation, increasing precipitation and further cooling of the atmosphere (Fagan 2000: 23–29).

Onset of the LIA was occasioned by a sequence of catastrophes throughout the fourteenth century and may have been triggered by several other events, among them increase in drifting Atlantic pack ice. The Great Famine of 1315–1322, fourteenth century's first major disaster, started with bad weather in the spring of 1315, setting a course of calamities that had continued to strike much of Europe. The famine was marked by extreme levels of starvation, disease, crime and infanticide. Millions of dead bodies had strewn the transition toward the LIA.

Barely recovered from the Great Famine, the bubonic plague, commonly known as the Black Death, struck Europe and the Near East over the period 1346–1353. The epidemic is thought to have originated in central Asia, carried to Europe by fleas on black rats that boarded merchant ships sailing from the Black Sea area to Constantinople, Genoa, Venice and Marseilles. Up to one half of the population of Europe and the Near East may have died in the plague, due to crowding and poor sanitation, with losses of life in cities near coastlines even higher (Fagan 2000: 223–244; 80–84). Romanesque streetscapes on the Continent and in England were a major accessory to the spread of the plague: most of the streets were narrow, crooked lanes through which open sewage was flowing unabashedly, with adjacent homes overcrowded and poorly ventilated (Deaux 1969: 122–3).

Whereas at the turn of the fourteenth century the populations of main urban centers of England, and what is now Germany, northern Italy, Belgium and France, by some accounts amounted to just over 3 million people, by the century's end these urban populations may have totaled less than 2 million (Cesaretti et al. 2016). Conservative estimates had put Europe's population at 73.5 million before the plague, and at 50 million at the end of the fourteenth century (Russell 1972). England may have lost up to 40% of its population, thought to have declined from about 7 million before the plague, down to as little as 4 million in 1400 (Russell 1948: 220–223). Population levels began to recover through the mid-fifteenth century, gaining momentum in the early sixteenth century. Resurgence of the plague had occurred periodically, sweeping through the Continent's different regions till the beginning of the nineteenth century but already by the turn of the seventeenth century Europe's population may have reached around 80 million.

The world population, by the turn of the seventeenth century, was between 550 and 580 million. North America's population, mainly the Aztec of Mesoamerica, and First Nations and the Inuit north of the Gulf of Mexico at that time accounted to about 20 million, with negligible European settler population. By comparison, toward the end of the nineteenth century, the world population was at 1.7 billion, Europe had 400 million people, and North America comprising Mexico, the United States and Canada, 94 million.

In Europe, the early calamities of the LIA had contributed to series of serfs' revolts and peasant uprisings that had flared toward the end of the fourteenth century in numerous locations throughout the Continent and intensified through the fifteenth and sixteenth centuries. The violence often came on the heels of failing crops and lack of other sources of subsistence, due to inclement weather which had led to short but repeated outbreaks of famine in western Europe during the seventeenth century and onward to the end of the LIA. In the view of Wolfgang Behringer modern European urbanization had been triggered, precisely, by considerable cooling that had forced people in the countryside, unprotected from bad weather as well as from marauding vagabonds and robbers, to seek shelter in cities (Behringer 2010: 57–59; 252–3).

The increase of population in Europe between the sixteenth and nineteenth centuries was largely due to urban growth. With the abolishment of serfdom, rapid urbanization mainly due to migration of peasants and freed serfs into cities had commenced, resulting in worsening gap between the rich and the poor in cities, between secular authority and the Church, and between various ethnoreligious and ethno-lingual groups. Somewhat in parallel to the hardship triggered by harsh climate, at the turn of the nineteenth century much of urban Europe and North America had embarked on a pattern of population change known as *demographic transition*, whereby—following extended migration into cities, along with advancing improvements in medicine and public health—lower death- and birthrates, had prompted accelerating growth as well as gradual aging of population (Caldwell 2006: 301–320).

The steady population increase in cities throughout central and western Europe during eighteenth century's Industrial Revolution had transformed onto commonplace occurrence the urban phenomenon of the crowd, and with it yet another, allied aspect of urban unfolding: the stranger. While in the Middle Ages crowds were usually seen only during major markets, carnivals or religious occasions such as pilgrimages, and a stranger in town would be usually instantly identified and singled out, the newly emerging daily spectacle of the urban crowd in cities of early modernity had resulted in the majority of urban dwellers becoming strangers to one another. Pertinent is a remark by René Descartes about his own human environment in the city of mid-seventeenth century. In a letter to his contemporary, Jean-Louis Guez de Balzac, Descartes uses the simile of a city walk to express his bliss of anonymity as a stranger within the crowd, along with what appears a measure of contempt:

> I can walk out each day in the bustle of the crowds with as much freedom and ease as you have in your paths, and I pay no more attention to the people I meet than I would to the trees in your woods or the animals that graze there. (Gaukroger 1995: 188)

By the end of the eighteenth century, the juxtaposition between the crowd and the individual had yielded two kinds of an urbanized stranger who stood out *from* the crowd: the flâneur, mainly in Paris, and the dandy, in Paris and London. The flâneur was an observer of the crowd, variably described by adjectives ranging from a lazy idler to a passionate onlooker. The dandy, on the other hand, had been a connoisseur of fashion through profligate clothing or demeanor, close to what might be called today, a metrosexual. Both largely male types had later spread to other European cities. The two urbanite types, the flâneur and the dandy, were not necessarily mutually exclusive, and they could be best characterized by their attitude toward the urban crowd. The transition from the flâneur onto the dandy is one from a bystander passively or fervidly observing the crowd, onto an extravagant enthusiast drawing the looks of the crowd upon himself. From the vantage points of both the flâneur and the dandy, the urban space is formed by the crowd and has little meaning without the crowd: the flâneur strives to be the observer of the crowd; the dandy strives to be the subject of observation by the crowd.

It is perhaps no overstatement that the flâneur and the dandy underwent a major shift at the turn of the nineteenth century with the advent of the shielded arcade where people-watching was now diverted from the open-air streetscape. It was due to the frequently inclement weather, characteristic of the LIA, that the covered glazed arcades had provided new public habitat and constituted a novel aspect of city-form responsive to bad weather, poor street maintenance and growing population at the end of this extended climatic aberration.

The world's first shopping arcades had emerged during the late eighteenth century. In response to the increasingly severe weather toward the end of the LIA, and the mounting urban populations, joint entrepreneurship between land tycoons and industrial builders had led to the construction of the first vast department store, the Great Guest Court in St. Petersburg, over the period 1757–1785. Some 2000 kilometers to the west, the Covered Market of Oxford had its initial core construction built during the years 1772–1773.

In Paris, the first indoor shopping gallery opened at the Palais Royal in 1786, followed by the Passage Feydau in 1790–1791 and the Passage du Caire in 1799. The first arcade with a glass rooftop was the Passage des Panoramas, opened in 1800 (Fierro 2003: 36). Other Parisian glazed-over arcades arose during the first half of the nineteenth century, posing sharp contrast to the city's unpaved streets with no sidewalks, where raw sewage mixed with mud was running down or was pooled in lagoons of stench and sludge. Elsewhere through Paris sewage was carried in open gutters as underground sewers could not sustain the city's rising population of over half-a-million people at the turn of the nineteenth century. To preserve their contrasting elegance, the arcades had mud-cleaning salons at their entrances ensuring that no soiled footwear would enter their sheltered world. As a connoisseur of fashion, the dandy would find the arcade a haven for safeguarding his appearance and asserting his own distinctiveness from the crowd.

1.3 Urbanization of the Late LIA and Its Cerebral Impact: Crowd as the Form of Space

Extreme volcanic activity in the southern hemisphere in 1815, releasing megatons of slowly spreading ash throughout the earth's atmosphere, had caused especially brutal weather the following year. Severe reduction in solar radiation reaching the earth's surface had led to a drop in average temperatures by almost one degree, triggering crop failures and food shortages in large parts of western Europe and the East Coast of North America (Post 1977: 12–13).

Dearth of cereals for livestock resulted also in the first large-scale dysfunction in urban transport. Lack of feed for animals had an immediate impact on land-transportation of the time, which was mostly horse-driven. The emerging urban transportation problem became the motivation that had led Karl Drais, chief forester of the Duchy of Baden, in the south of Germany, to invent horseless conveyance. The year 1816, dubbed as the 'Year Without a Summer,' was also the year Drais ushered the era of mechanized personal transport by his invention of the *draisine*. An inline two-wheeled vehicle, with the front wheel and handlebar hinged to allow steering, the *draisine* was propelled by the rider pushing along the ground with the feet as in regular walking or running. Although extremely uncomfortable to ride, the *draisine* had become a fad, particularly among the young generation in circles seeking extravagance:

> Some media portrayals flattered the Draisine; but many more pilloried or satirized it. The popular term that stuck to it was 'hobby horse' or 'dandy-horse' [...] generally seen as the domain of dandies and privileged university students (Schiller et al. 2010: 60)

Some 40 years after Drais' invention, the first pedal-equipped velocipede, developed and patented by the Parisian blacksmith Pierre Michaux, became the forerunner of the modern bicycle. Between 1857 and 1871 the Michaux Company had mass-produced the velocipede (Herlihi 2004: 420), an appealing human-propelled vehicle for the streets of Paris, now newly paved under a massive streetscape alteration scheme.

Urban transportation was the lesser reason why in 1853 baron Georges-Eugène Haussmann, the prefect of the *departement de Seine*, launched a large-scale reconstruction of central Paris. In 1851 Louis-Napoleon Bonaparte, the original Napoleon's nephew, seized power from the government of the Second Republic of which he was the president and declared himself Emperor Napoleon III. Haussmann was installed with the official charge to allow to clean up Paris, to unify the city with its suburbs and to beautify it. Severely inadequate sanitary infrastructure in the capital had brought repeated outbreaks of epidemics, particularly in the overcrowded medieval core of central Paris. Haussmann was tasked with the redesign of the heavily populated, poverty-stricken and crime-infested urban core of Paris. Napoleon III, fearful of the violence emanating from central Paris, wanted also to make it his legacy to end the periodic returns of pestilence to the city. The narrow streets void of sunlight and air circulation, the urban setting of repeated riots and armed violence against the authorities, were demolished in Haussmann's redesign that was to tackle the entire range of emerging urban issues, from street safety, security, surveillance and crowd control to air circulation and public health, to architectural design and traffic.

Haussmann, described as a dandy, undoubtedly projected his own extravagance upon the ideal city he wanted Paris to become through his designs (Touttain 1971: 30–37). With most of the impoverished residents relocated or otherwise displaced, broad tree-lined boulevards were carved through what were previously people's homes and neighborhoods. The crooked lanes and winding streets of medieval Paris vanished, giving way to fashionable promenades, not unlike London's Regent Street, where middle-class Parisians came to survey Haussmann's controversial urban spectacle and to become spectacle themselves.

With Haussmann's redesign of Paris, modernity's surrogate for chance encounter in the urban streetscape became the gaze: the crowd inspected by an observer or anomalous individual scrutinized by the crowd. The built environment of modernity and the planned streetscape geared to safety and security become thus instrumental in facilitating the circumstances of the gaze. An alienating relation between strangers within the crowd, the gaze has reinforced the urban streetscape as a medium of surveillance.

It was this Haussmannian ambiance within which the relationship between the crowd and the individual observing it in the French capital had been memorably detected and put into a literary discourse during the 1860s by Baudelaire:

> The crowd is his element, as the air is that of birds and water of fishes. His passion and his profession are to become one flesh with the crowd. For the perfect flâneur, for the passionate spectator, it is an immense joy to set up house in the heart of the multitude [...] (Baudelaire 1864/1964)

Baudelaire's lyrical description of the flâneur and the crowd would be next to impossible to make for a Romanesque streetscape. The paradigmatic medieval street was crooked, and the streetscape was often an incoherent mesh of streets and small lanes. In Haussmann's redesign of Paris, a guiding principle had been the facility of crowd control and the ability for surveillance. It is the latter feature that would have energized the urban phenomenon of the flâneur and the dandy in the streetscape. Haussmann and Baudelaire stand on the extreme opposite ends of urbanist outlooks, but the free-spending Baudelaire, who in a short time went through all his inheritance, was known as a dandy too.

The inadvertent result of configuration of streets and edifices during the Romanesque was an urban continuum of chance encounter between people. The Romanesque urban space, due to its confinement within city walls, was typified by high round-the-clock residential density. The resultant feature of the medieval city was the continuing opportunity for a sporadic open-air human encounter during daytime. This has changed with intensifying urbanization, and particularly with the wholesale, deliberate demolition of city walls in larger centers during the nineteenth century. Unchecked by a physical boundary of the walls suburban sprawl had begun, and overall population *density* within city limits had declined, even though population within city limits had increased. The chance for acquaintances to randomly cross path had diminished, and human encounter in the city had become formalized to transactions between strangers.

In historic and prehistoric communities, from Neolithic settlements to the towns and hillforts of the Iron Age through to Romanesque cities, encounter was usually between humans known well to one another; it was the obvious, unremarkable and unspoken ongoing occurrence. With intensifying urbanization encounter between acquaintances in the city became increasingly rare. Clarity in monumental streetscapes of the kind of Nash's elegant Regent Street or Haussmann's fashionable redesign of Paris had replaced the enigma of medieval street network, and the occasion of commonality in an acquaintance encounter had transformed into prevalence of a stranger's gaze. At the turn of the twentieth century, streetscape clarity and harmony had become an urban design standard of the City Beautiful urbanist movement in North America (Bassnett 2007), and by the mid-twentieth century a canon of socialist realism in urban architecture of the Soviet Union and its satellite countries (Hatherley 2016: 9–28, 103–115). By then urban estrangement had turned into urban resentment.

The American City Beautiful movement of the early 1900s, and socialist realism in urban architecture in the USSR and its satellites of the 1950s and 1960s had both advocated monumental civic designs of envisaged urban harmony and beauty. America and the Soviet Union, twentieth century's political and moral adversaries, through their respective mainstream architects and urban planners, saw civic beauty as an architectonic framework that was to mold urban inhabitants to standards of moral and civic virtue. That virtue was to be aligned with the respective worldviews of the two powers, which happened to be entirely opposed to one another. Architectural determinism had guided the City Beautiful movement (Burgess 1997), much as it was the backbone of socialist realism in architecture and urban planning (Crowley 2010).

Setting aside the architectural determinism that had guided the two urbanist ideologies of the capitalist West and the communist East, a caustic view of the American City Beautiful movement and of the Soviet planned city points to an underlying modern urban feature, shared by both: the dread of the crowd. In America the City Beautiful was driven by middle-class fear of the impoverished, the homeless or the delinquent roaming the streets (Bess 1990); in the Soviet Union the planned city, such as Magnitogorsk, had become the prototype of the urban assembly line and surveillance of the crowd, a "living space under stranger's gaze" (Kotkin 1995: 157–197).

In the geographic and ideological space between America and the Soviet Union, Benjamin pondered the human condition within the modern city-form. Decades after the *Parisian Scenes*' urban poetry in *The Flowers of Evil* and 100 years after the first arcades were introduced in the French capital, it was the cognizance of estrangement in the city that contextualized Benjamin's psychoanalytic survey of urban edifices and spaces of Paris and Berlin. Advancing the premise that indeterminate but unrelenting mutual impact between contemporaneous built environment, on the one hand, and humans within it, on the other hand, ought to be counted as a force behind the evolution of urban civilizations, Benjamin came to predicate estrangement as a perpetuating force molding twentieth century's city-form.

Throughout the twentieth century Benjamin's notion, nuanced in his *The Work of Art in the Age of Mechanical Reproduction* (1936/2008), had been validated by housing projects or suburban subdivisions often exuding alienation as a self-perpetuating anomaly, feeding upon itself, a sentiment nascent from gargantuan ideal-city concepts of automation and uniformity, such as the Radiant City of Le Corbusier or Broadacre City of Frank Lloyd Wright. Ingrained in projects of the two towering architects and urbanists of the twentieth century, alienation ought to be seen also as part and parcel of their documented personality disorder (Evenson 1979: 199–219, 232–238; Friedland and Zellman 2006: 251–340, 543–61). Other instances of twentieth-century architect-planners with cognitive or emotional aberrations abound, their projects sometimes focused on surveillance and control (Hagman 2010: 110–112; Lahiji 2011). In their footsteps went the offspring of urban modernity, architects or developers of infamous housing projects and cookie-cutter subdivisions seldom paused to reflect on the consequences of their work for clients they never met in person. Humans growing up and living in urban projects of the 1950s and 1960s would come to know little of authenticity of an original, in a world where the copy or the assembly line had become the standard (Fig. 1.2).

Yet the notion of surveillance and with it, alienation, seems to have been disguised in the paradigm of the ideal city much earlier than modernity. Mimicking the perceived perfection of the universe and implying a patronizing Grand Designer, ideal-city concepts from Plato to modernity have echoed the belief that city-form ought to reflect cosmic faultlessness.

In Plato's Myth of Er (*Republic* X, 615b), seven inner celestial spheres of ancient cosmology are described as world whorls, a variant of ancient beliefs about the universe. Seven is also the number of walls encircling the archaic city of Ecbatana in Persia (James and van der Sluijs 2008), there are seven inner zones of water and land in Plato's Atlantis (Golding 1975), repeated in the imaginary design of the City of the Sun by Tommaso

Fig. 1.2 Pruitt-Igoe housing project, St. Louis, 1951–1972/1976, by award-winning architect, Minoru Yamasaki. (Source: U.S. Department of Housing and Urban Development Office of Policy Development and Research)

Campanella (1602/1981: 27) or in the seven circular boulevards of the Happy Colony of Robert Pemberton and the Garden City of Ebenezer Howard in the second half of the nineteenth century (Emmons 2015). In more than one way, Howard's Garden City is the culmination of Neo-Platonic ideal-city notions. Municipal services in Howard's urban concept are worked into minute detail but fire halls and police stations are strikingly missing in his Garden City (Howard 1902/1946). The wide boulevards ensuring transparency in streetscapes, along with the amicable urban environment of the Garden City, ostensibly, would eliminate crime and accidents, but evidently also, would facilitate surveillance across the Garden City.

Against the archetypal myth of the ideal city, there is much merit to Benjamin's discernment of urban authenticity, manifest as the overt appearance of the "unconscious of the collective." To Benjamin, the street since antiquity, and throughout history to our own times, has been a frequent medium by way of which minds have generated images and thoughts. On this view, there are thus two elements that had set in motion the interaction of minds and their built environment: the one is the myth of the ideal city; the other is the authentic, unembellished space of the city street.

In this sense contrast between a mythical ideal world and the unpretentious urban reality also meaningfully echoes Niccolo Machiavelli who, a century before Descartes, Kepler and Galileo, may have determined the gist of modern thought as a drive to juxtapose authoritarian civic imagery against genuine reality. Alluding to the mythical ideal city-state of Atlantis in Plato's *Republic*, Machiavelli wrote in chapter 15 of his short treatise, *The Prince*:

> Men have imagined republics and principalities that never really existed at all. Yet the way men live is so far removed from the way they ought to live that anyone who abandons what is for what should be pursues his downfall rather than his preservation [...].

Benjamin's apposition of myths of urban utopia, against the urban reality of everyday street, is also consistent with the fundamental thrust of modern thought toward critical observation and examination. Furthermore, in envisaging a single hybrid of city-form and minds within it, Benjamin makes a corollary statement of philosophical nature. Putting forward the notion of a relentless *reciprocal* feedback between the street, as a perceptual emblem of contemporaneous built environment, on the one hand, and human minds within it, on the other hand, Benjamin envisages the myth of the ideal city as constructively conducive to the evolution of the built environment. It is the interaction between the myth of the ideal city and the incessant urban experience within streetscapes that constitutes forceful dynamics behind the unfolding of city-form as well as urban civilizations.

Myths and rituals on the Italian peninsula had transformed into the custom of sacred ceremonies in the founding of cities already in the Etruscan period (*c.* 700 BCE) and continued to be the practice for almost a millennium, throughout much of the Roman Republic and Empire. Fabled founders of cities in Greek mythology are often deified heroes, and within newly founded Greek colonies streets had often been named after deities (Calame 2003: 52–85). The archetypal myth of the ideal city falls within this very same pattern. Benjamin had broached the question of mind-city interaction, pointing to the query of the occurrence and situation of humans within their built environments as delineating the history of civilization. By drawing on Carl Jung's notion of the collective unconscious, Benjamin observes "elements of ur-history" wedded in an epochal feedback pattern, producing in their contact with authentic urban environments a newly shared ideal, or common elements in the visioning of a utopia. The shared elements of a community's image of a utopia stamp their mark upon contemporaneous configuration of urban objects, "from permanent buildings to ephemeral fashions" (Benjamin 1935/1969).

In observing the condition of individuals within the twentieth century's urban environments Benjamin had focused on environ/*mental* interaction with the city's edifices, and voids between or within them. The mind-city feedback that Benjamin conceives of consists of bilateral and interminable metamorphosis of ideal-city myth and authentic city-form. City-form mutates when the myth of an ideal city is brought to bear upon it, while the myth of the ideal city transforms itself in the same feedback process when new features of city-form emerge. The contemporary city is the result of just such feedback loop, ongoing through history.

The philosophical significance of the *Arcades Project* is the emerging grasp of modernity through the notion of contemporaneous urban space as a stage in the mind-city feedback progression. Urban space, as Benjamin characterizes it, is distinct from the space of nineteenth century's textbook physics: it is a fluid, dynamic and idiosyncratic medium. The meaning of urban space emerges through its interaction with the crowd:

> *Streets* are the dwelling place of the collective. The collective is an eternally restless, eternally moving essence that, among the facades of buildings endures, experiences, learns, and senses as much as individuals in the protection of their four walls. For this collective the shiny enameled store signs are as good and even better a wall decoration as a salon painting is for the bourgeoisie. Walls with the "*defense d'afficher*" are its writing desk, newspapers are its libraries, letterboxes its bronzes, benches its bedroom furniture – and café terraces the balcony from which it looks down on its domestic concerns after work is done. (Benjamin 1933/1999: 423)

To the extent that a streetscape and its fixed features are deliberately designed, they invite usage of an implicitly declared category: sidewalks are for walking but not for cycling or skateboarding, for example; benches are for sitting, but not for sleeping. Yet regardless of what the builder or designer intent has been in producing public space, it is the undisciplined crowd who through its dynamic and entirely unpredictable modes of usage of public space and its features determines multiplicity of its evolving meanings. The urban space of modernity, in particular, is not at all an objective intermediary between physical points, as is its Newtonian counterpart, the physical space. To the extent that urban voids, such as streets, and objects within them, such as street furniture, are conduits to human movement, they attain meaning through ever-changing modes of usage by the crowd. It was possibly in this sense that observation of crowds within the urban space had led Baudelaire to assert: "By modernity I mean the transitory, the fugitive, the contingent" (Baudelaire 1864/1964:13).

To Benjamin, reflecting on Baudelaire, urban space is not a passive and immutable receptor of movement and change, but it is a vital and invigorating constituent in mind-city reciprocity. Urban space and objects within it interact with the crowd as dynamic fluents (Mueller 2015: 9–11). Within such interpretation, the notion of urban space extends the concept of space (and time) introduced by Immanuel Kant in the eighteenth century. To Kant, the Euclidian space of daily experience or Newtonian physics is the form of perceiving. Introducing a new paradigm of space, as well as time, Kant's contention was that mind is incapable of perceiving the world as it truly is, except only through its forms of perceiving which the mind projects upon the thing-in-itself, a transcendent object inaccessible to mind directly. Space and time are mind's a priori forms of perceiving, and the world in space and time as we perceive it is the synthesis of the thing-in-itself *and* space and time; the thing-in-itself is beyond space and time.

Kant, a geography professor at the University of Königsberg, could hardly have been unaware of a folkish conundrum circulating in his native city during his boyhood. It was the question how to traverse the seven bridges of Königsberg in a single path so that each bridge could be crossed once and only once. The puzzle ended up tackled by Leonhard Euler, the Swiss mathematician who, proving that the task is impossible to accomplish, had inadvertently founded graph theory (Gribkovskaia et al. 2007). The connection of space and movement with the lived experience of the geometry of a streetscape could have been also some early impetus to Kant's own revolutionary advent of space and time as constructs of mind. But whereas to Kant physical, Newtonian space is the form of perceiving, to Benjamin urban space is *formed* by the crowd. The crowd is, as it were, the form of modernity's *urban* space.

1.4 Myths of the Garden and the Citadel in the Emergence of Conceptual Dispositions

Starting with Kierkegaard and Nietzsche the notion of the crowd has connoted usually negative association of *passive* herd-like mentality. Such view was further reinforced in later Existentialist literature by José Ortega y Gasset in *The Revolt of the Masses* (1932/1957). Ortega's focal notion is the *señorito satisfecho*, 'Mr. Satisfied,' the basic component of the crowd who knows his narrow task to be performed at his job, but not much else beyond. In Chapter XII entitled 'The Barbarism of "Specialisation"' Ortega introduces a paragraph on scientific expertise that in time came to be known for its notoriety:

> The specialist serves as a striking concrete example of the species, making clear to us the radical nature of the novelty. For, previously, men could be divided simply into the learned and the ignorant, those more or less the one, and those more or less the other. But your specialist cannot be brought in under either of these two categories. He is not learned, for he is formally ignorant of all that does not enter into his speciality; but neither is he ignorant, because he is "a scientist," and "knows" very well his own tiny portion of the universe. We shall have to say that he is a learned ignoramus, which is a very serious matter, as it implies that he is a person who is ignorant, not in the fashion of the ignorant man, but with all the petulance of one who is learned in his own special line. (Ortega 1932/1957: 112)

In what is now commonly referred to as the Ortega Hypothesis, the controversial proposition states that progress in science, in fact, is made in advances of small steps by multitudes of ordinary scientists, rather than by leaps of the singular scientific genius. In the ensuing discussion, cultural and historical context of scientific discovery has been often put forward in argumentation for or against (e.g., Cole and Cole 1972).

But passivity assigned to the crowd, from Kierkegaard to Ortega, is most certainly not an attribute assigned to the crowd by Benjamin. Furthermore, communities bound by geography and some common history have always exerted also an active impact upon their environment, usually affirmative to the community itself. In their most rudimentary modes of impact upon the environment, humans since prehistory have projected their own gender, the simplest and the very basic demographic characteristic, upon their material culture and upon forms of their built environments. Usually, inquiries into the built environment have ignored such considerations.

The notion of the crowd probably came to the cognizance of humans sometimes early in the Neolithic Demographic Transition (NDT), commonly known as the Neolithic Agricultural Revolution. Marking the transition from nomadic hunter-gatherer nomadic communities to permanent settlements, people stayed in large groups for longer periods of time than earlier on. Guiding the NDT was climate, namely, the end of the Ice Age and the onset of the Holocene. Increased food intake led to rapid population growth throughout Eurasia, accompanied by the founding of Neolithic villages and hamlets. The most rudimentary of demographic traits, gender, had stamped its mark on the first built environments and had continued to exert its subtle impact upon urban environments throughout history, and to this day.

Western city form has been, primarily, the product of a masculine myth, transmuted in its most vivid paradigm into the Platonic Ideal City. The claim that the evolution of ideas throughout history went hand in hand with the evolution of city-form ensues from the founding proposition of perpetual feedback between minds and their built environment (Akkerman 2016: 6–7). This environ/*mental* process had sprung to life early in the dawn of civilization, and the same process had continued to shape thought-doctrines as well as evolution of the city. In the ongoing historical process of urban-cerebral progression, the myth of the Ideal City gives rise to actual city-form, followed by cultural and intellectual change that occurs within the changed urban environment. Newly emergent outlooks, in turn, yield also new perceptions of the Ideal City. This epochal feedback pattern has constituted the imposition of a primarily masculine attribute of the Ideal City upon the evolution of thought as well as upon the advances of city-form. Philosophical urbanism, as the very argument for the existence of this process, projects this historical exposition upon concerns with the contemporary evolution of ideas and the city.

The masculine myth of the Ideal City appears to be deeply intertwined with a feminine counterpart: The Garden. In Carl Jung's psychoanalytic theory, too, a man carries within the ingrained makeup of his personality an archetypal feminine side, the Anima, while a woman carries within her own subliminal traits a correspondingly opposite masculine imprint, the Animus, both archetypes of the collective unconscious (Jung 1953/1966: 188–211). It is edifying, in this regard, to juxtapose Jung's archetype of the mother, along with the Earth Mother goddess as its epitome, with the conjecture by the eighteenth-century philosopher, Giambattista Vico, on the prehistoric emergence of cities. Jung's introduction of archetypes of the collective unconscious and Vico's account in his *Scienza*

Nuova (1744/2002: IV, 982) on the origin of cities are a forceful note on the Biblical narrative of the Garden and the Citadel. Vico was the first to suggest that prehistoric cities arose as hunters, fathers of families, forming pacts to face off external attack as well as to suppress internal mutiny, became the first overlords, the *builders* of citadels and forts. On this view, a myth of the Garden had evolved as environmental paradigm of the Earth Mother, possibly during the Upper Paleolithic, *c.* 50,000–12,000 BP. Symbolizing fertility, the female gatherers of fruits and vegetables became the Garden's emissaries.

The Myth of the Garden, thus, is primeval while that of the Citadel is at the founding of early civilization. The Myth of the Citadel arises along with the first appearance of citadels and forts as a masculine paradigm. Following Vico's suggestion, then, the intrepid Citadel becomes the masculine consort of the seraphic Garden, and this may have occurred with the first prehistoric settlements sometime after 15,000 BP.

Primal sacredness in the Bible, in contrast with the sinful Citadel, is conferred upon the Garden. The Garden in Genesis expresses purity in a primordial interaction with the environment and innocence in relations between humans.

> And out of the ground the Lord God formed every beast of the field, and every fowl of the air; and brought them unto the man to see what he would call them; and whatsoever the man would call every living creature, that was to be the name thereof. [...] Therefore shall a man leave his father and his mother, and shall cleave unto his wife, and they shall be one flesh. And they were both naked, the man and his wife, and were not ashamed. (*Genesis* II: 19, 24–25)

But in the Bible there is also an ominous relation between the Garden *and* the Citadel: the need for shelter—for a citadel or a city—arises with the expulsion from the Garden. Yet the expulsion is nothing but an inevitable outcome of innocent altruism that leads to the first sin: the sharing in the Garden's Forbidden Fruit. At the end of selflessness in the Garden, the dread of human existence emerges through the discovery of reason and through the rise and decline of the Citadel. As a narrative sequence in the Bible, the Garden and the Citadel are inseparable, constituting a sequence in space and time.

In Jungian psychoanalysis too archetypes and myths are reciprocally interwoven within the structure of the human mind. Emerging during the early history of humanity, both in Biblical scorn of the Tower of Babel and in Greek mythology and philosophy, the Myth of the Citadel unfolds into the myth of the Ideal City. The Hebrew Bible presents the building of the Tower of Babel as a precursor to the human concept of the Ideal City, in contrast to a divine scheme, and it associates the citadel's construction with reason:

> Now the whole earth had one language and few words. And as men migrated from the east, they found a plain in the land of Shinar and settled there. And they said to one another, 'Come, let us make bricks, and burn them thoroughly.' And they had brick for stone, and bitumen for mortar. Then they said, 'Come, let us build ourselves a city, and a tower with its top in the heavens, and let us make a name for ourselves, lest we be scattered abroad upon the face of the whole earth.' And the Lord came down to see the city and the tower, which the sons of men had built. And the Lord said, 'Behold, they are one people, and they have all one language;' [...] and from there the LORD scattered them abroad over the face of all the earth. (*Genesis* XI: 3–4)

Paul Lampl (1968: 7–12) has shown that the account of the Tower of Babel is in sharp contrast with most Near-Eastern mythologies, in which cities are created by gods and with classical Greece, where cities (or even streets) are dedicated to gods. In the Hebrew Bible, furthermore, contemporaneous cities are not only manmade but also drawn in a genealogy of sin:

> Cain said to Abel his brother, 'Let us go out to the field.' And when they were in the field, Cain rose up against his brother Abel, and killed him. [...] Then Cain went away from the presence of the LORD, and dwelt in the land of Nod, east of Eden. Cain knew his wife, and she conceived and bore Enoch; and he built a city, and called the name of the city after the name of his son, Enoch. (*Genesis* IV: 8, 16–17)

The myths of the Garden and the Citadel could be the primordial environmental parallel, respectively, to the cults of Dionysus and Apollo at the Delphi temple in the northern Mediterranean. Friedrich Nietzsche, in his very first book, *The Birth of Tragedy*, shows the Greek tragedy as arising from two unconscious impulses which he called the Dionysian and the Apollonian drives or premonitions. The Dionysian drive corresponds to spontaneity and to the flow of unadulterated, subjective experience, and it is expressed in the temporal arts such as music and dance. The Apollonian instinct is behind the structure, order and restriction of visual arts, and it is expressed primarily in sculpture and architecture:

> It is by those two art-sponsoring deities, Apollo and Dionysus, that we are made to recognize the tremendous split, as regards both origins and objectives, between the plastic, Apollonian arts and the non-visual art of music inspired by Dionysus. The two creative tendencies developed alongside one another, usually in fierce opposition, each by its taunts forcing the other to more energetic production, both perpetuating in a discordant concord that agon which the term *art* but feebly denominates: until at last, by … an Hellenic act of will, the pair accepted the yoke of marriage and, in this condition, begot Attic tragedy, which exhibits the salient features of both parents. (Nietzsche 1871/1956: 19)

Emerging from nature's wilderness, Dionysus, the god of wine, agriculture and fertility, carries the earmark of uninhibited passion, while Apollo—the god of prophecy and the patron of cities and city walls—is associated with control and the imposition of limits. Nietzsche's juxtaposition of the Dionysian and Apollonian impulses as gender types is no accident. The prehistoric emergence of the myths of the Garden and the Citadel also seems to inform, respectively, the prevailing urban cultures of ancient Israel and classical Greece. While Israelite settlements emphasize open space as a link to the rural and nomadic background from which they grew, Greek cities accentuate buildings, a testimony to analytical ability and intellectual prowess. In more than a figurative sense, the urban culture of ancient Israel implies a *time*-bound disposition, implicit also in its agricultural tradition and religion, while the distinct characteristics of Greek temples and planned colonies, a reflection of Greek mythology and philosophy, are their *spatial* features (Akkerman 2002).

1.5 Jung's Castle in the Forest: Gender in Human Niche Construction

In 1913, at the age of 38, Jung had a spell of what appeared to have been a series of psychotic experiences. He recorded his chimeric reveries in notes that he transcribed into a large red leather-bound folder. Analyzing his own notes, intermittently over 16 years, Jung finally produced the three-part *Red Book* (1913/2009), or what came to be known also as the *Liber Novus*.

The second chapter of the book's second part, *Liber Secundus*, is entitled 'The Castle in the Forest.' Jung's trance account starts with description of "a small old castle" surrounded by a swamp in "a dark forest" (p. 220). The servant of the castle's owner, an old scholarly man, offers Jung a room for the night. Appearing at the door of Jung's room and heading to the foot of his bed is a beautiful young woman, the old man's daughter and personification of Salome, New Testament's notorious Jewish princess, infamous for ordering the severed head of John the Baptist to be brought to her on a plate (*Mark* 6: 17–29 and *Josephus* XVIII, 5: 4). A passionate dialogue with Jung concludes with the young woman's words, "I thank you, dear friend, and I bring you greetings from Salome," her shape then dissolving into darkness. "Where she stood something shadowy lies – it is a profusion of red roses" (p. 225).

Against Jung's suggestive fantasy looms famous contemporaneous painting of a beautiful girl with a crown of Jewish stars atop red roses in her hair by the Czech Art Nouveau illustrator, Alphonse Mucha. *Princess Hyacinth* was created by Mucha in 1911 as an advertisement poster for eponymous fairy tale ballet and pantomime played in the Czech National Theatre in Prague in the same year. Mucha had created a color print, *Salome*, in 1897 while in France, and in 1904 the French illustrator Gustav-Adolf Mossa had created also a *Salome*, symbolic painting of a *femme fatale* with vampire tendencies, a "woman-child surrounded by blooming roses, each containing the bleeding head of John the Baptist" (Neginsky 2013: 78). Mucha's *Princess Hyacinth* of 1911, seems to be yet another insinuation of Salome, this time inspired by Mossa's painting.

The premiere show of *Princezna Hyacinta* was in Prague on September 3, 1911. The third international Psychoanalytic Congress at which Jung was elected the first president of the International Psychoanalytic Society, was held about 150 kilometers from the Czech border (marking the boundary between Germany and the Austro-Hungarian Empire), at the Erbprinz Hotel in Weimar, Germany, on September 21–22, 1911. It is quite likely that Mucha's poster was displayed at the hotel and throughout the town, and that it would come to Jung's sight there or elsewhere. Whether or not Mucha actually intended to picture Salome as the princess Hyacinth in his poster may not matter, if in the image on the poster Jung saw the subject of his own obsession. Jung's infatuation with the persona of Salome has been seen in context of larger public fascination with this New Testament female character, at the turn of the twentieth century (Anthony 2018: 2–3, 96–97). The prejudice and bigotry in Jung's younger years have been either documented (Sherry 2010: 52–58) or passed over in silence, and his questionable relations with at least some Jewish women in his life has also been laid bare (Sells 2017: 143–168).

The Salome reverie, within the context conjectured here, is possibly a confirmation not only of Jung's human lapses in the first half of his life but, more significantly, also a testimony of archetypal projection of gender imprints upon the built environment. Two and a half thousand years before Jung, Plato depicted the mythical origins of the fabled ideal city of Atlantis in a fashion not entirely unlike Jung's description of the Castle in the Forest. In the dialogue *Critias* Plato purports to give a description of the island of Atlantis, founded by the god of sea, Poseidon, some "9000 years" before Solon, a legislator who preceded Plato by two centuries (*Timaeus* 23d-e). In Plato's myth, at the center of the island, an old man lived on a mountain:

> He was called Evenor and he lived with his wife, Leucippe. They had just the one child, a daughter called Cleito. When the girl reached the age of marriage, both her mother and her father died, but Poseidon, who had come to desire her, made her his concubine. He gave the hill where she lived secure defences by breaking it off the surrounding land and creating increasingly large concentric rings, alternate of land and water around it. Two of the rings were of land, three of water, and he made them equidistant from the centre, as if he had taken the middle of the island as the pivot of a lathe. (*Critias* 117d)

James and van der Sluijs (2008) have shown that the plan of Atlantis, described in *Critias*, is analogous to the ancient geocentric view of the cosmos. Neolithic carvings of concentric rings, known as cup-and-ring marks, as well as concentric stone circles, have been extant in many places throughout the world, including the Mediterranean. The cup-and-ring marks, as well as the concentric stone circles, are most likely depicting stars rotating around the north celestial pole (Castleden 1987). Arguably then, the source of the myth of Atlantis is in the rotating sky, an archetypal origin to an ideal built form on earth, as implied also by Plato's 'Myth or Er' at the very end of *The Republic* (Akkerman 2014).

Jung's enigmatic account in 'The Castle in the Forest,' with a scene of "dark swamp water, and a small old castle standing at its center," continues with the story of an old man and his young daughter. This makes the parallel with Plato's Atlantis almost complete, as Jung's object of desire, similar to Poseidon's, is a young woman. The archetypal evidence is in the similarity of both stories, and it begs an archaic context to a more comprehensive consideration.

Jung had returned to a cryptic environmental motif about 1931 when he wrote *Mind and Earth*, a controversial essay that, in suggesting that different places impact mind and behavior in different ways, displays modern affinity with Montesquieu's environmental determinism. Jung's first critics, Karl Popper chief among them (Popper 1963: 37–38), have variously accused him of unscientific bias since his propositions are not falsifiable, and thus ought to be relegated to the category of myth. Early criticisms of Jung have later met considerable opposition, largely based on the observation that stringent standards applicable to traditional physical science, would deem many other fields, such as parts of biology or ecology, as pseudoscience. Unhelpful to Jung's legacy, however, were some of his early statements, easily seen as bigoted or otherwise disdainful. Yet it is the more recent look at Jung's *Mind and Earth* that has led some to revisit its message:

> Jungians may have been reluctant to explore the relation between mind and earth, in part because of the disquieting controversy over Jung's early anti-Semitism and the Nazi motto *Blut and Boden*, or *Blood and Soil*. Despite these associations, however, the ecological crisis, among other things, forces us to reopen the discussion about mind and earth. (Tacey 2009)

Taken together, Jung's two short essays could be seen an esoteric reflection binding together the million-year evolutionary segment from *Homo erectus*, the first species able to control fire, to *Homo sapiens*, the only species to use human language, in a progression extended to our very days (Percival 1993).

The extended part of this continuing progression of humanity starts with the millennia-long transition from primeval human-built shelters of the Middle Paleolithic, c. 300,000 BP, into the first human settlements of the late Upper Paleolithic, c. 15,000 BP. This transition is at the commencement of the Neolithic period starting with the NDT and continuing through the rise of civilization to the present day. Adducing Jung's collective unconscious, Walter Benjamin addresses the mind-city progression within the latter period:

> In the dream in which every epoch sees in images the epoch that follows, the latter appears wedded to elements of ur-history […]. Its experiences, which have their storage place in the unconscious of the collective, produce, in their interpenetration with the new, the *utopia* that has left its trace behind in a thousand configurations of life, from permanent buildings to ephemeral fashions. (Benjamin 1972: 46–47, quoted in Buck-Morrs 1990: 114)

The building of shelter began with the need to moderate the effects of climate, controlling the immediate environment as a compensation for the gradual shedding of fur among some of the hominid, the great apes. The endeavor of hominid and human shelter construction corresponds to the notion of extended phenotype in ecology, introduced by Richard Dawkins, whereby organisms are recognized to modify their habitat niches through their activities and choices, while the niches sometimes yield a reverse impact upon the genetic makeup of organisms inhabiting them (Dawkins 1978). Dawkins' notion of the extended phenotype has led to the recognition of impact that biological niche construction carries upon the direction in the evolution of species when organisms interact not only with their immediate habitat niche but also with each other's habitat niches.

John Odling-Smee has expanded this notion to *human* niche construction where he includes environmental complexity, encompassing urban habitat, within what he calls *cultural inheritance* of humans. Cultural inheritance so understood is in an ongoing interaction with the genetic inheritance of contemporary humans (Odling-Smee et al. 2003: 239–281). His sweeping assertion that "in modern humans, cultural inheritance is so powerful that in many cases it no longer interacts with genetic inheritance at all, but overrules it" (Odling-Smee et al. 2003: 246), may seem radical when considered within the scale of less than a few human generations. But over extended historic time, such statement is not overbearing at all, as is also arguably the evolutionary case of humans. Similar to the extended phenotype in biological species, the harnessing of fire, perhaps 1.3 mya, with the first human construction of shelter, about 300 kya, had facilitated not only adaptability to a range of climatic conditions but had also extended gradual impact upon the genetic makeup of human beings.

Groups of Paleolithic people, especially those who lived in cold climates, initially found shelter in caves or later also in tents constructed of branches and animal skins. Primitive shelters built at first by anatomically modern humans endured only short periods of time, days, weeks or months. Over millennia, however, human-made structures became more environmentally resistant. Through the construction of first built environments, human use of fire and hides had led to the cognition of communal structure. John Gowlett suggests that a large portion of nighttime activity was spent around the fire, contributing to the evolution of language and social interaction among individuals (Gowlett 2016). In a later development, the harnessing of fire had led to the construction of first hearths or other fire enclosures such as circles of stones, further fostering language among community that would gather round (Chazan 2017). The control of fire became a useful skill among early humans that may have led to social stratification and to differentiation of communal roles through the separation of groups engaged in cooking from groups engaged in hunting (Wrangham 2009: 121–141).

Similar to fire, preparing and curing of skins into hides, too, had led to the construction of artificial shelter offering protection from the elements of nature as well as from animals. Vegetation, long bones of animals and hides were adapted into coverings as protection from cold, heat and rain, especially as humans migrated to colder regions (Wenke 1984: 148). Stone slabs were used to surround firepits, sometimes building structures over them using mammoth bones for frames and animal hide for walls and roofs, which then provided for stability, security and comfort in communities of early modern humans (Redman 1999: 53–80).

The oldest known human habitation site, Terra Amata, southern France, dated to about 230 kya, had consisted of primeval huts on a beach and traces of domestication of fire, the earliest known in Europe (de Lumley 2010: 211–213). While the men of Terra Amata hunted elephant, rhinoceros and deer, the women gathered fruits and vegetables in the nearer vicinity and maintained and controlled the fire. This was the likely pattern of human existence during the Middle to Upper Paleolithic throughout large parts of Africa and Eurasia.

The extended phenotype notion of Dawkins expanded by Odling-Smee onto human niche construction, encompasses also human habitat transformation from the Upper Paleolithic to the Neolithic. Further, it ought to be examined also against the notion, *cognitive niche*, initially proposed by Tooby and DeVore (1987). Applying the word "niche" as a metaphor rather than a reference to a small, geographically confined locale, Tooby and DeVore point to the colonization of ecosystems by humans as a unique stage in the earth's environmental history, seen as a coevolutionary relationship between humans and their environment:

> Humans have made a unique and major innovation in this coevolutionary relationship – an innovation that is our most distinctive characteristic. The defenses of plant and animal prey species can be circumvented by "surprise" attacks, attacks that consist of many novel and discrete manipulations, evolutionarily unprecedented actions that prey defenses are not equipped to repel. Goal-oriented actions by humans, shaped to suit the particular situation, constitute surprise attacks. We accomplish this by conceptually abstracting

> from a situation a model of what manipulations are necessary to achieve proximate goals that correlate with fitness. These highly orchestrated and intricate situation-specific sequences of behavior are cognitively organized. The core of our zoological distinctiveness is our entry into this cognitive niche. (Tooby and DeVore 1987)

The authors' allusion to a scheme or a plan by humans colonizing their environment, while also compensating for their physiological inferiority against game, for example, seems to extend to shelter and settlement construction. More recently, Lorenzo Magnani introduced the notion 'cognitive niche construction' as an interactive "process in which concepts, ideas, and thoughts are projected onto external structures," creating a feedback pattern between the mind and its environment: "Once concepts and thoughts are externalized and projected, new chances and ways of inferring come up from the blend" (Magnani 2008).

Magnani's notion applies to contemporaneous built environment much as it does to the natural environment of early humans. The discovery, harnessing and adaptation of fire and hides by early humans is a case in point and a prime example of mutual-impact progression in cognitive niche construction as delineating the evolution of humans. In an example of evolutionary change of the human body, many anthropologists believe that the discovery and control of fire by *Homo erectus*, more than a million years ago, had contributed to the gradual loss of human body hair (Medler 2011). The deliberate wearing of hides by early *modern* humans, the Homo sapiens, about 0.5 mya possibly also contributed to this process.

Controlling fire or the wearing of clothes are not confined to a constant *geographical* niche but they can be considered "cognitive niche construction" leading to the loss of body hair as a specific impact. Furthermore, the comparative lack of body hair in women, as opposed to men, has been explained by the likely early habit of human females, extending possibly over hundreds of thousands of years, to wear hides in excess, not only for protection but also as ornamentation, such as in sexual signaling, similar to jewelry (Aldhouse-Green 2004).

Impact upon early built environments by human females extends this consideration by the proposition that "female bonding may have been a more powerful force in human evolution than is sometimes supposed" (Dunbar 1996: 150). Of some significance ought to observation of female bonding in primates closest to humans:

> [chimpanzee] female tends to socialize only with a subset of other females, some of whom may be relatives from the natal group. Female chimpanzees tend to travel with one another, and they may repel strange females trying to enter their group. This behaviour implies a close female community. (McKee et al. 2016: 101)

If females formed the sedentary or domestic core of the early groups of humans, language as a bond within these groups, may have evolved first among the human females at spontaneous gatherings. Such gatherings could not occur at treetops but, more likely, in open spaces during clement weather. Evolution of human language was probably aided by the emergence of open space as public space. Complex human behavior, including vocalized language, had evolved during the Middle Stone Age in Africa, 280 kya–50 kya (Bickerton 2003; Perreault and Mathew 2012), and the emergence of space designated for gatherings, particularly in open-air, may have occurred in low latitude climates through much of the year.

Perceived bonding and interaction among human females within early built environments could further explain the transformation of geographic void in the Neolithic settlement, onto a commons intended for deliberate gatherings. The unfolding of geographic void onto a public space, as a feminine trait within early human habitat, is thus consistent with views stressing female capacity in the evolution of language. Such a view does not necessarily preclude the conventional outlook on early language, founded on belief in the imperative of cooperation in men's activities, such as hunting.

Unfolding of early public space may have evolved also along with male-female bonding rituals through music and dance during the Upper Paleolithic. The progression of music and dance into public ritual would have required specific open space that would have been likely actively searched for and designated as public ritual place. First musical instrument, a hollow bone, has been dated to 60,000 BP (Arensburg et al. 1989), but vocal communications between mothers and infants (Dissanayake 2000) or imitation of rhythmic noises produced by rattling seeds or pounding roots into a meal were much earlier likely musical sources (Morley 2013: 16–31). Dance is thought to have originated at the start of the Upper Paleolithic (Garfinkel 2010), possibly turning into part of fertility rituals where singing and dancing individuals "were to demonstrate their physical and psychological fitness to prospective mates" (Miller 2000). Dunbar points to feelings of euphoria produced through music, song and dance, possibly as a result of secretion of endorphin hormones in the brain which occurs after vigorous exercise (Dunbar 2012). Articulation of open-air space, designated for public rituals, could be then also dated to the Upper Paleolithic, between 50,000 and 15,000 BP.

Some later prehistoric communal open spaces as cultic places are extant still today. Concentric rings of monumental menhirs, standing stones, on the empty Salisbury Plain in Wiltshire, England, were being laid out between c. 3000 and 2000 BCE in a vast megalithic earthwork, known today as Stonehenge. England's Stonehenge or the nearby Avebury are open-air areas laboriously turned during the late Neolithic into designated public spaces, their cultic aspect apparent. Some evidence suggests that both Stonehenge and Avebury were used, among other things, for renewal rituals, which likely included fertility rites (Meaden 1999: 27–36, 102–130).

Focus on open space as a communal asset may have also occurred already with the communal fire-pit during the Upper Paleolithic, or earlier. Early emergence of the public space, as a place of ritual dance, for example, may have been initially linked to nonverbal communication and to the procurement of language later on. Open communal space would be also quite distinct from the more private shelters in its vicinity. But whereas human-built shelters, even if only primitive huts, constituted a tangible, largely vertical, alteration within the human environment of the Paleolithic, the mutual configuration of such structures and their geographic distribution had been subtle and less perceptible. Focus on the horizontal aspects of the built environment, facilitating movement and mutual access were the gist of more abstract thinking millennia later, with the onset of the Neolithic and the NDT. Distribution of dwellings within the first settlements of the Neolithic had expressed a level of contemplation, giving rise to the urban theme that had led to the emergence and planning of first cities during the Bronze Age.

The planning and design of human habitat through recorded history became the expression of horizontal considerations of movement within the built environment, while in its primordial trait, architecture came to accentuate the static, vertical object of the built form. In a generic setting of a communal human habitat, the distinction between architecture and urban planning and design could be characterized by the tension between *volume*, a masculine projection represented by vertical structures, and *void*, a feminine imprint embodied by open spaces within a settlement. The transformation onto city-form of streetscapes, parks and gardens, and urban squares, on the one hand, and edifices, from single-family homes to cathedrals and high-rises, on the other hand, does not change the premise of the city's foundational makeup as the projection of void and volume, the archaic archetypes of femininity and masculinity.

Bibliography

Akkerman, Abraham. 2002. Altruism and Egoism: The Garden and the Citadel. *The Structurist* 41/42: 65–72.

———. 2013. Gender Myth and the Mind-City Composite: From Plato's Atlantis to Walter Benjamin's Philosophical Urbanism. *GeoJournal* 78 (4): 727–741.

———. 2014. Platonic Myth and Urban Space: City-Form as an Allegory. *University of Toronto Quarterly* 83 (4): 757–779.

———. 2016. *Phenomenology of the Winter City: Myth in the Rise and Decline of Built Environments*. Dordrecht/New York: Springer.

Aldhouse-Green, Miranda. 2004. Crowning Glories: Languages of Hair in Later Prehistoric Europe. *Proceedings of the Prehistoric Society* 70: 299–325.

Anthony, Maggy. 2018. *Salome's Embrace: The Jungian Women*. New York: Routledge.

Arensburg, B., A.M. Tillier, B. Vandermeersch, H. Duday, L.A. Schepartz, and Y. Rak. 1989. A Middle Palaeolithic Human Hyoid Bone. *Nature* 338 (6218): 758–776.

Bassnett, Sarah. 2007. Visuality and the Emergence of City Planning in Early Twentieth-Century Toronto and Montreal. *Journal of the Society for the Study of Architecture in Canada* 32 (1): 21–38.

Baudelaire, Charles. 1857/2006. *The Flowers of Evil*. Middletown: Wesleyan University Press.

———. 1864/1964. *Painter of Modern Life and Other Essays*, ed. Jonathan Mayne, 22–40. London: Phaidon.

Behringer, Wolfgang. 2010. *A Cultural History of Climate*. Cambridge: Polity Press.

Benjamin, Walter. 1933/1999. *The Arcades Project*. Cambridge, MA: Harvard University Press.

———. 1935/1969. Paris: Capital of the Nineteenth Century. *Perspecta* 12 (1969): 163–172.

———. 1935/1983. *Charles Baudelaire: A Lyrical Poet in the Era of High Capitalism*. London: Verso.

———. 1936/2008. *The Work of Art in the Age of Mechanical Reproduction*. New York/London: Penguin Books.

———. 1940/1999. *The Arcades Project*. Trans. Howard Eiland and Kevin McLaughlin, Rolf Tiedemann (ed.). Cambridge, MA/London: Belknap Press of Harvard University.

Bergson, Henri. 1903/1912. *Introduction to Metaphysics*. New York: G.P. Putnam & Sons.

Bess, Philip H. 1990. Big Plans, Divine Details. *Inland Architect* 34: 56–63.

Bickerton, D. 2003. Symbol and Structure: A Comprehensive Framework for Language Evolution. In *Language Evolution*, ed. M.H. Christiansen and S. Kirby, 77–93. Oxford: Oxford University Press.

Bocquet-Appel, Jean-Pierre. 2008. Explaining the Neolithic Demographic Transition. In *The Neolithic Demographic Transition and Its Consequences*, ed. Jean-Pierre Bocquet-Appel and Ofer Bar-Yosef, 35–55. New York: Springer.

Buck-Morrs, Susan. 1990. *The Dialectics of Seeing: Walter Benjamin and the Arcades Project*. Cambridge: MIT Press.

Burgess, Patricia. 1997. The Expert's Vision: The Role of Design in the Historical Development of City Planning. *Journal of Architectural and Planning Research* 14 (2): 91–106.

Calame, Claude. 2003. *Myth and History in Ancient Greece: The Symbolic Creation of a Colony*. Princeton: Princeton University Press.

Caldwell, John C. 2006. *Demographic Transition Theory*. New York: Springer.

Campanella, Tomasso. 1602/1981. Civitas solis. In *The City of the Sun: A Poetical Dialogue by Brother Tommaso Campanella*, ed. Daniel J. Donno. Berkeley: University of California Press.

Castleden, Rodney. 1987. *The Stonehenge People: An Exploration of Life in Neolithic Britain 4700–2000 BC*. London and New York: Methuen.

Cesaretti R., J. Lobo, L.M.A. Bettencourt, S.G. Ortman, and M.E. Smith. 2016. Population-Area Relationship for Medieval European Cities. *PLoS One* 11 (10): e0162678. https://doi.org/10.1371/journal.pone.0162678. http://journals.plos.org/plosone/article?id=10.1371/journal.pone.0162678

Chazan, Michael. 2017. Toward a Long Prehistory of Fire. *Current Anthropology* 58 (S16): S351–S359.

Cole, Jonathan R., and Stephen Cole. 1972. The Ortega Hypothesis. *Science* 178 (4059): 368–375.

Crowley, David. 2010. Paris or Moscow? Warsaw Architects and the Image of the Modern City in the 1950s. In *Imagining the West in Eastern Europe and the Soviet Union*, ed. György Péteri, 105–130. Pittsburgh: University of Pittsburgh Press.

Dawkins, Richard. 1978. Replicator Selection and the Extended Phenotype. *Zeitschrift für Tierpsychologie* 47 (1): 61–76.

de Lumley, Henry. 2010. *La Grande Histoire des premiers hommes europeens*. Paris: Odile Jacob.

Deaux, George. 1969. *The Black Death, 1347*. London: Hamish Hamilton.

Diamond, Jared. 1999. *Guns, Germs, and Steel: The Fates of Human Societies*. New York/London: Norton.

Dissanayake, E. 2000. Antecedents of the Temporal Arts in Early Mother-Infant Interaction. In *The Origins of Music*, ed. Nils Wallin, Bjorn Merker, and Steven Brown, 389–410. Cambridge, MA: MIT Press.

Duffy, F. 1974. Office Design and Organizations: 1. Theoretical Basis. *Environment and Planning B* 1 (1): 105–118.

Dunbar, Robin I.M. 1996. *Grooming, Gossip and the Evolution of Language*, 149–150. London: Faber and Faber.

———. 2012. On the Evolutionary Function of Song and Dance. In *Music, Language and Human Evolution*, ed. N. Bannan and S. Mithen, 201–214. Oxford: Oxford University Press.

Emmons, Paul. 2015. Reading Between the Lines: The Dark Side of Ebenezer Howard's *Garden Cities of To-Morrow*. In *Architecture's Appeal: How Theory Informs Architectural Praxis*, ed. Marc J. Neveu and Negin Djavaherian, 33–46. London/New York: Routledge.

Evenson, Norma. 1979. *Paris: A Century of Change, 1878–1978*. New Haven/London: Yale University Press.

Fagan, Brian. 2000. *The Little Ice Age: How Climate Made History 1300–1850*. New York: Basic Books.

Ferris, D.S. 2008. *The Cambridge Introduction to Walter Benjamin*. Cambridge/New York: Cambridge University Press.

Fierro, Alfred. 2003. *La Vie des Parisiens sous Napoleon*. Saint-Cloud: Soteca Napoleon I.

Foucault, Michel. 1978/2009. *Security, Territory, Population: Lectures at the College De France, 1977–78*, ed. Michel Senellart. New York: St Martins Press.

Friedland, Roger, and Harold Zellman. 2006. *The Fellowship: The Untold Story of Frank Lloyd Wright and the Taliesin Fellowship*. New York: Regan.

Garfinkel, Yosef. 2010. Dance in Prehistoric Europe. *Documenta Praehistorica* 37: 205–214.

Gaukroger, Stephen. 1995. *Descartes: An Intellectual Biography*. Oxford/New York: Oxford University Press.

Golding, Naomi H. 1975. Plato as City Planner. *Arethusa* 8: 359–371.

Gowlett, J.A.J. 2016. The Discovery of Fire by Humans: A Long and Convoluted Process. *Philosophical Transactions of the Royal Society B*. 371 (1696): 20150164.

Gribkovskaia, Irina., Øyvind Halskau Sr., and Gilbert Laporte. 2007. The Bridges of Königsberg: A Historical Perspective. *Networks* 49 (3): 199–203.

Hagman, George. 2010. *The Artist's Mind: A Psychoanalytical Perspective on Creativity, Modern Art and Modern Artists*. London/New York: Routledge.

Hatherley, Owen. 2016. *Landscapes of Communism: A History Through Buildings*. New York: New Press.

Herlihi, David V. 2004. *Bicycle: The History*. New Haven: Yale University Press.

Howard, Ebenezer. 1902/1946. *Garden Cities of To-morrow: A Peaceful Path to a Real Reform*. London: Faber and Faber.

James, Peter, and Marinus Anthony van der Sluijs. 2008. Ziggurats, Colors, and Planets: Rawlinson Revisited. *Journal of Cuneiform Studies* 60: 57–79.

Johnson-Hanks, Jennifer. 2008. Demographic Transitions and Modernity. *Annual Review of Anthropology* 37: 301–315.

Josephus, Flavius. 1c CE/2006. *Jewish Antiquities*. Ware: Wordsworth.

Jung, C.G. 1913/2009. *The Red Book: Liber Novus*, ed. S. Shamdasani. New York: W. W. Norton.

———. 1916/2009. *Liber Novus: A Reader's Edition*, ed. and with an Introduction by Sonu Shamdasani. New York/London: W.W. Norton.

———. 1953/1966. The Relation Between the Ego and the Unconscious. In *Collected Works of C.G. Jung*, ed. Carl G. Jung, vol. 7, 2nd ed. Princeton: Princeton University Press.

Kierkegaard, Søren. 1859. *The Crowd is Untruth: On the Dedication to "That Single Individual"*. Trans. Charles K. Bellinger. https://oregonstate.edu/instruct/phl201/modules/Philosophers/Kierkegaard/kierkegaard_the_crowd_is_untruth.html. Accessed 16 May 2017.

Kotkin, Stephen. 1995. *Magnetic Mountain: Stalinism as a Civilization*. Berkeley/London: University of California Press.

La Mettrie, Julien Offray de. 1748/1994. *Man a Machine and Man a Plant*. Trans. Richard A. Watson and Maya Rybalka, Introduction and Notes by Justin Leiber. Indianapolis: Hackett Publishing Co.

Lahiji, Nadir. 2011. Must Architecture Be Defended? The Critique of Violence and Autoimmunity. In *Architecture and Violence*, ed. Bechir Kenzan, 199–218. New York: Actar.

Lampl, Paul 1968. *Cities and Planning in the Ancient Near East*. New York: George Braziller.

Leslie, Esther. 2007. *Walter Benjamin*. London: Reaktion.

Logan, F. Donald. 2002. *A History of the Church in the Middle Ages*. London/New York: Routledge.

Lowenthal, David. 1975. Past Time, Present Place: Landscape and Memory. *Geographical Review* 65 (1): 1–36.

———. 1994. Author's Response (to David N. Livingstone's 'Classics in Human Geography Revisited'). *Progress in Human Geography* 18 (2): 213–215.

Lunn, Eugene. 1982. *Marxism and Modernism: An Historical Study of Lukács, Brecht, Benjamin, and Adorno*. Berkeley/London: University of California Press.

Magnani, Lorenzo. 2008. Chances, Affordances, Niche Construction, In *Knowledge-Based Intelligent Information and Engineering Systems: 12th International Conference*, ed. Ignac Lovrek, Robert J. Howlett, and L.C. Jain, KES 2008, Zagreb, Croatia, September 2008. Proceedings, Part II.

McKee, Jeffrey K., Frank E. Poirier, and W. Scott Mcgraw. 2016. *Understanding Human Evolution*. New York: Routledge.

Meaden, Terence. 1999. *The Secrets of the Avebury Stones: Britain's Greatest Megalithic Temple*. Berkeley: North Atlantic Books.

Medler, Michael J. 2011. Speculations About the Effects of Fire and Lava Flows on Human Evolution. *Fire Ecology* 7 (1): 13–23.

Miller, G. 2000. Evolution of Human Music Through Sexual Selection. In *The Origins of Music*, ed. Nils Wallin, Bjorn Merker, and Steven Brown, 329–360, 389–410. Cambridge, MA: Massachusetts Institute of Technology.

Montesquieu, Charles Louis de Secondat. 1749/1949. *The Spirit of the Laws*. English Trans. T. Nugent. New York: Hafner Publishing Co.

Morley, Iain. 2013. *The Prehistory of Music: Human Evolution, Archaeology, and the Origins of Musicality*. Oxford: Oxford University Press.

Morris, A.E.J. 2013. *History of Urban Form Before the Industrial Revolution*. Oxon/New York: Routledge.

Mueller, Erik T. 2015. *Commonsense Reasoning: An Event Calculus Based Approach*. New York/Amsterdam: Elsevier.

Neginsky, Rosina. 2013. *Salome: The Image of a Woman Who Never Was*. Newcastle Upon Tyne: Cambridge Scholars.

Nietzsche, Friedrich. 1871/1956. *The Birth of Tragedy from the Spirit of Music*. New York: Random House.

———. 1886/1997. *Beyond Good and Evil: Prelude to a Philosophy of the Future*. New York: Dover Publications.

Odling-Smee, John, Kevin Laland, and Marcus Feldman. 2003. *Niche Construction: The Neglected Process in Evolution*. Princeton: Princeton University Press.

Ortega y Gasset, José. 1932/1957. *The Revolt of the Masses*. New York: Norton.

Panofsky, Erwin. 1957. *Gothic Architecture and Scholasticism*. New York: Meridian.

Pearce, Trevor. 2010. From 'Circumstances' to 'Environment': Herbert Spencer and the Origins of the Idea of Organism-Environment Interaction. *Studies in History and Philosophy of Biological and Biomedical Sciences* 41 (3): 241–252.

Peet, Richard. 1985. The Social Origins of Environmental Determinism. *Annals of the Association of American Geographers* 75 (3): 309–333.

Percival, R.S. 1993. Is Jung's Theory of Archetypes Compatible with Neo-Darwinism and Sociobiology? *Journal of Social and Evolutionary Systems* 16 (4): 459–487.

Perreault, Charles, and Sarah Mathew. 2012. Dating the Origin of Language Using Phonemic Diversity. *PLoS One* 7 (4): 1–8.

Poe, Edgar Allan. 1840/2004. Man of the Crowd. In *The Collected Tales and Poems of Edgar Allan Poe*, 207–2013. Ware: Wordsworth.

Popper, Karl. 1963. *Conjectures and Refutations: The Growth of Scientific Knowledge*. London: Routledge & Kegan Paul.

Post, John D. 1977. *The Last Great Subsistence Crisis in the Western World*. Baltimore: Johns Hopkins University Press.

Redman, Charles L. 1999. *Human Impact on Ancient Environments.* Tucson: University of Arizona Press.

Russell, Josiah Cox. 1948. *British Medieval Population.* Albuquerque: University of New Mexico Press.

———. 1972. Population in Europe: 500–1500. In *The Fontana Economic History of Europe*, The Middle Ages, ed. Carlo M. Cipolla, vol. I, 25–71. Glasgow: Collins/Fontana.

Sauer, Carl O. 1925. The Morphology of Landscape. *University of California Publications in Geography* 2 (2): 19–53.

Schiller, Preston L., Eric Bruun, and Jeffrey R. Kenworth. 2010. *An Introduction to Sustainable Transportation: Policy, Planning and Implementation.* London/Washington, DC: Earthscan.

Sells, Angela M. 2017. *Sabina Spielrein: The Woman and the Myth.* Albany: SUNY Press.

Sherry, J. 2010. *Carl Gustav Jung: Avant-Garde Conservative.* New York: Palgrave Macmillan.

Spencer, Herbert. 1864. *The Principles of Biology, Volume 1.* Edinburgh: Williams and Norgate.

Tacey, David. 2009. Mind and Earth: Psychic Influence Beneath the Surface. *Jung Journal* 3 (2): 15–32.

Tooby, J., and I. DeVore. 1987. The Reconstruction of Hominid Evolution Through Strategic Modeling. In *The Evolution of Human Behavior: Primate Models*, ed. W.G. Kinzey, 183–237. Albany: SUNY Press.

Touttain, Pierre-André. 1971. *Haussmann: artisan du Second Empire, créateur du Paris moderne.* Paris: Librairie Gründ.

Villa, Paola. 1983. *Terra Amata and the Middle Pleistocene Archaeological Record of Southern France.* Berkeley: University of California Press.

Wenke, Robert J. 1984. *Patterns in Prehistory: Humankind's First Three Million Years.* Oxford: Oxford University Press.

Wrangham, Richard. 2009. *Catching Fire: How Cooking Made Us Human.* London: Profile Books.

CHAPTER 2

Sky and Gender Myths in the Founding of Early Built Environments

Introduction and Summary

The Neolithic Demographic Transition (NDT) was preceded and conditioned by an earlier demographic event, the migration from the Horn of Africa to Eurasia by *Homo erectus*, and later by anatomically modern humans, the *Homo sapiens*. The migration waves 'Out of Africa' concluded about 70,000 BP dispersing humans throughout southern Eurasia and later to Australasia. The most perceptible demographic earmark of humans, their gender, was also the foundational premise of the mind-environment interaction. Initial stage in the interaction between mind and the environment was effected by the one-directional impact in the discernment of the human body, the object that is absolutely immediate to the senses of the perceiving subject. Against the consummately closest object, one's own body stands the object unreachably distant: the sky. The very first universal synthesis between the minds of humans and their environment was the projection of gender upon perception of the environment, the earth and the sky.

The first such projection led to human perspicacity of the sky, the North Star in particular, as a masculine entity, and to the discernment of the earth as a female. A bright North Star, as a fixed beacon in the sky, was the object pursued by herds of *Homo erectus* prompting them northward. The North Star, around which the nightly firmament revolves, became the epitome of masculine prowess, strength and certainty at night, corresponding to daylight's sun, the patriarchal provider and protector, or

A. Akkerman, *Philosophical Urbanism*,
https://doi.org/10.1007/978-3-030-29085-6_2

alternatively, the tyrannical ruler and unyielding magistrate. The earth, life-giving and whimsical, came to represent the Great Mother, a complement to the masculine celestial deity.

Such outlook helps explain a conundrum related to Upper Paleolithic Venus figurines in Eurasia, believed to represent the Great Mother goddess. The geographic pattern in the distribution throughout Eurasia of the Upper Paleolithic sites where the Venus figurines were found, constitutes a geographic band along the 50th parallel. This geographic belt, throughout Eurasia, would be under severely cold conditions, and the riddle rests in the question how, or why, did the Venus Figurines find their way to such harsh environment. The answer likely rests in the conjecture that the Venuses symbolize, indeed, a mother goddess. The feminine art of Paleolithic Venus statuettes in Siberia and in eastern and central Europe seems to complement mythology of a masculine North Star, a counterpart to a prevailing earthbound myth of the Earth Mother. It is at the geographic latitude of the Venus finds where the North Star would be easily visible in the sky, not too close to the horizon *or* to zenith.

Sites of the Paleolithic Venus figurines are all within a Eurasian band along the 50th parallel, approximately, a vast region in the western part of which, millennia later, cup-and-ring marks were carved into rocks at hundreds of sites during the late Neolithic and the Bronze Age. The cup marks are usually found near or at plateaus, highlands and seashores where the north celestial pole could be observed unobscured as a pivot of the revolving nightly sky. The cup marks, as well as the prehistoric earth- and stonework roundels, suggest the forming of a renewal parable, the Eternal Return, during the late Neolithic when the North Star was no longer easily discernible in the nightly sky. At the absence of a bright North Star in the nightly sky, the dark void at the northern celestial pole had served a deified feminine feature around which all stars revolved.

To the early human observer of the sky, gender projection upon the northern celestial pole had resolved the riddle of nightly firmament rotation through mythical axle of the world, the *Axis mundi*, bringing the sky and the earth into a union. Neolithic and Bronze Age cup-and-ring marks, along with the lithic or ditched roundels, were expressions of the human attempt to explain the workings of the sky and the seasons. The Upper Paleolithic Venuses and the Neolithic cup-and-ring carvings, two distinct types of prehistoric stone art, point also to the origins of the mythical belief in cyclical renewal, rebirth and recurrence.

Axis mundi and the Eternal Return had become the archetypal notions through which humanity came to perceive the environment. During the Neolithic, these two environmental imprints of the mind projected upon the *built* environment had yielded, on the one hand, the *round* open space, a designed void and the public, open-air hub of a community, and on the other hand, the vertical volume, the edifice serving as a shelter and containing domestic space. Benjamin's own struggle with the Eternal Return in his *Arcades Project* suggests an archetypal power of this concept. Similarly, too, the notion of the *center*, implicit in an axle, has been foundational through the history of the city and the history of architecture, as Christian Norberg-Schulz has shown.

2.1 Early Anthropomorphic Projections Upon Geographic Space

Emergence among the Late Upper Paleolithic people of designed void as a primordial public space can be conjectured from some of the later Neolithic and Bronze Age stonework, extant till today. As the ensuing discussion suggests, concentric circular earthworks from the Neolithic, and the concurrent appearance of concentric rings in stonework, known as cup-and-ring marks, imply such a linkage through what appear to be fundamentally feminine features of designed void. Not only the Stonehenge monumental site in England but also round enclosures in central Europe are believed to have been venues for renewal rites pursued during the Neolithic. Some of these sites are believed to have been scenes of sexual rituals, precursors of divine marriage enactment of the later Bronze and the Iron Ages, the *hieros gamos*. These public sites of Neolithic communities came to be also the grounds of budding commons and primordial civic space.

The open Neolithic public space has its source in myth, millennia earlier, in the Upper Paleolithic art of Venus figurines, female statuettes whose shared features are their exaggerated pelvic area and breasts. To C.G. Jung the Venus figurines represent the Earth Mother (Jung and Kerenyi 1951/1977, CW 9i, ¶ 312). The Paleolithic Venuses are carved from soft stone, bone or sculpted and fired as a ceramic artifact, and to many anthropologists too they represent a myth of the Great Mother or the Earth Mother. The Paleolithic Venuses are dated to between 40,000 BP and 15,000 BP, that is, much of the prehistoric time period known as Upper Paleolithic. The Upper Paleolithic, c. 50,000 BP–12,000 BP, heralded the coming end of the Stone Age and of the Ice Age, as it also overlapped the latter half of the Last

Glacial Period, 115,000 BP–12,000 BP. The Ice Age, or the Pleistocene, was the geological period, 2.5 mya–12,000 BP, whose latter portion, c 300,000 BP–70,000 BP, included the human migration waves 'Out of Africa' east and northward.

The human migration waves 'Out of Africa' were countered in Eurasia by expanding continental-scale ice sheets and Alpine glaciers produced during glacial episodes, the last such episode being the Last Glacial Period (LGP). It was during the cold LGP during which early populations of *Homo sapiens* spread through Eurasia, largely south of the ice sheets or in between icefields. It is remarkable that the figurines consistently point to geographic origins of very cold climate. The geographic locations of the Venus finds form a conspicuous pattern of a discernible geographic corridor, roughly along the 50th parallel. Most of the Paleolithic Venus sites within this corridor are in central Europe, while the remainder is sparingly distributed in central Russia and Siberia. The figurines seem to have ceased from being produced within this geographic corridor during the Neolithic and later, when increasingly sophisticated sculpting of Venuses appears in more southerly latitudes, closer to the Tropic of Cancer (about 23°26′ N).

Rather than originating in temperate regions nearer the equator, the Venuses were probably manufactured near the sites of their finds, along the 50th parallel, often north of it, surrounding time and latitude associated with the frigid Last Glacial Maximum (LGM), spanning a period when Earth's ice sheets were at their greatest range, c. 31,000–16,000 BP. Many of the extant Paleolithic Venus statuettes are attributed to a toolmaking culture referred to as the Gravettian, c. 30,000–20,000 BP, just prior to the largest extent of glaciation in the western part of Siberia and in Europe (Barr and Clark 2011). Growth of the ice sheets reached maximum range in Europe c. 22,000–18,000 BP and in western Siberia approximately 18,000–17,000 BP. Eastern Siberia, including the easternmost site of the Venus figurines at Lake Baikal, was an exception as it was not covered by an ice sheet but was interspersed with icefields covering mountain ranges (Niessen et al. 2013; Margold et al. 2016). Regardless of the actual proximity of the ice, however, it is clear that the Venus sites had been in very cold regions. Notably, the Venuses had been produced many millennia prior to the advent of pictographic tablets that preceded cuneiform script and formal counting in the temperate regions of Egypt, the Levant and Mesopotamia.

Even the unlikely assumption, that the figurines were produced at some southerly latitudes of Eurasia and carried northward, does not diminish the enigma of their location pattern (Table 2.1). Placement of the Upper Paleolithic Venuses, along the corridor of high geographic latitude, and in

Table 2.1 Upper Paleolithic sites of Venus figurines, 40,000–13,000 BP

Loc	*Venus site*	*Region*	*Modern boundaries*	*Est. production time*	*Latitude*	*Longitude*
1	Lake Baikal	S-C Siberia	South-Eastern Russia	24,000–15,000 BCE	53°30′N	108°0′E
2	Gagarino	Lipetskaya Region	Central Russia	22,000 to 18,000 BCE	52°42′N	39°09′E
3	Kostienki	Borshevo District	Voronezh Region, C. Russia	23,000–20,000 BCE	51°24′N	39°02′E
4	Zaraysk	South of Moscow	Central Russia	20,000 BCE	54°46′N	38°53′E
5	Avdeeva	Near city of Kursk	Kursk District, Central Russia	21,000–20,000 BCE	51°43′N	36°11′E
6	Petřkovice	Northern Moravia	Czech Republic	23,000 BCE	49°52′N	18°15′E
7	Parabita	Lecce, SE Italy	Italy	15,000 BCE	40°03′N	18°08′E
8	Moravany n. Vah	Near Piešťany	Western Slovakia	21,000 BCE	48°36′N	17°52′E
9	Dolní Věstonice	Southern Moravia	Czech Republic	29,000–25,000 BCE	48°53′N	16°38′E
10	Galgenberg	Lower Austria	Austria	28,000 BCE	48°27′N	15°36′E
11	Willendorf	Lower Austria	Austria	28,000–25,000 BCE	48°17′N	15°24′E
12	Savignano	Modena	Northern Italy	25,000–20,000 BCE	44°29′N	11°2′E
13	Schelklingen	Near Ulm, Swabia	Southern Germany	38,000 to 33,000 BCE	48°23′N	09°44′E
14	Engen	Konstanz	Southern Germany	11,000 to 15,000 BCE	47°51′N	08°46′E
15	Grimaldi Caves	Balzi Rossi	Northwest Italy	18,000 BCE	43°47′N	07°32′E
16	Gönnersdorf	Near Neuwied	West-Central Germany	13,000 to 10,000 BCE	50°26′N	07°28′E
17	Monruz	Near Neuchâtel	Switzerland	11,000 to 15,000 BCE	47°0′N	06°56′E
18	Cave of Lespugue	Pyrenees Mountains	SW France	24,000 to 22,000 BCE	43°25′N	01°30′E
19	Cave of Laussel	Dordogne Valley	SW France	23,000 BCE	45°0′N	0°40′E
20	Cave Brassempouy	Landes	SW France	23,000 BCE	43°38′N	0°42′W

the immediate vicinity of frozen-over areas, points to some momentous reason, universal to the people of Eurasia at these extremely inhospitable locations of a very cold climate. To people for whom fire was a treasure such universal reason could hardly be anything else but the sun during daytime and the permanent flare of a bright North Star in the nightly sky. The North Star, encircled by revolving stars, was the shared observational experience of all upright standing humans at open spaces and within a casual vertical field of view. For an upright standing human, a casual vertical field of view is the range formed by an arc between the horizon (0°) and approximately 50° above the horizon.

It is the auspicious feature of the Pole Star, that the angle at which the Pole Star appears to a viewer on earth, that is, its altitude above the horizon, equals the latitude of the viewer's location (Nicolson 1999: 38–39). In particular, then, the altitude of the North Star in the sky at any location equals the latitude of the location (Aveni 2001: 49–54). Up to about 50° northern latitude the North Star appears within an ordinary field of vision of any human being in a casual upright position. Humans in a vertical posture, who are at the latitude of about 50° N or less, need not significantly lift their heads in order for the Pole Star to come into their view.

Herein also lies the resolution to much of the enigma of the Venuses' location pattern. As a pivot star nearest to the north celestial pole, the North Star is seemingly at the center of nightly rounds made by revolving stars. Those stars that complete a full circle round the North Star, without setting under the horizon, are the circumpolar stars (Nicolson 1999: 39–40). The further north an observer moves the higher is the altitude in the sky of the North Star, and accordingly also, is higher the number of circumpolar stars. In southerly latitudes of the northern hemisphere, the North Star approaches the horizon, and most of the stars that can be observed are not circumpolar, as they rise above the horizon in the evening and set in the early morning. Any circumpolar stars in close angular vicinity to the North Star may not even be visible pending on the brightness of the North Star. Furthermore, in low latitudes, the North Star itself may not be visible at all, due to topography or high vegetation such as forests.

On the other hand, for example, at about 20° N, the North Star in Northern Polynesia, while almost always within a convenient view, just above the sea horizon, has almost no associated circumpolar stars to discern. In contradistinction, mythology of the Inuit, though involving stars and asterisms, does not employ the North Star at all (MacDonald 1998: 169–173; Dunne 2011: 252). In latitudes higher than 50° the pole star begins to approach the zenith, and in order to see the Pole Star, one must lift one's head. This becomes progressively strenuous with closer proximity to the pole where the Pole Star is exactly overhead and thus becomes also useless for navigation. At the latitude of the Inuit, virtually all stars are circumpolar, but to discern the North Star itself, almost at the zenith, a very inconvenient posture would be required for an extended time. Polaris for the Inuit is known as 'Nuuttuittuq' but in myth "the North Star [...] appears to have been unknown to the peoples of the North" (VMC 2014). Even in more contemporary Inuit legends, there is no mention of the North Star (MacDonald 1998: 59–62). In more southerly north-hemispheric mythologies the sky has chiefly the attributes of masculinity (e.g., Griffin-Pierce 1992: 174–195; Leeming 2010: 332). In the ancestral mythology of America's Great Plains First Nations, the North Star has usually played a masculine role (Dorsey 1997: 134–137). The North Star has a masculine attribute also in Taoist religion (Li and Wong 1994: 24).

2.2 North Stars of the Equinoctial Cycle: From Eurasian Migrations to the Venus Figurines

The Earth's celestial poles change in accordance to the oscillation of the Earth's axis, in a cycle of about 25,800 years, known as precession of the equinoxes. During this cycle, the swaying Earth's axis points to different sections of the sky. The incumbency of a North Star is thus entirely given by the oscillation of the Earth's axis during the equinoctial cycle. Incumbency of the North Star during the equinoctial cycle was likely of critical significance in the migratory waves of *Homo erectus* from Africa: the upright posture of *H. erectus* had facilitated casual observation of the revolution of the nightly sky about the North Star, and when the incumbent North Star was particularly bright, it would have served as a permanent and unchanging beacon in the sky.

The hypothesis that early humans had evolved in eastern Africa is supported by fossil analysis as well as some genetic evidence (Hammer et al. 2011). Groups of *H. erectus* began leaving Africa eastward and northward in multiple migratory waves starting probably some 1.2 million years ago. In Eurasia *H. erectus* diverged into the Neanderthal as well as into other species of the genus *Homo*, and into *H. sapiens*, to which all modern humans belong (Holt 2015). The divergence of the Neanderthal and archaic *H. sapiens* species from *H. erectus* has been estimated to have occurred between 800,000 BP and 400,000 BP (Endicott et al. 2010). In Africa the species *H. sapiens* had emerged between 400,000 BP and

200,000 BP, expanding sporadically through the Horn of Africa to Eurasia. Humans through the migration waves from Africa interbred with or replaced earlier populations of the genus *Homo* that became resident outside Africa earlier on, such as the Neanderthals. During the period between 125,000 BP and 70,000 BP groups of *H. sapiens* left Africa in last migration waves of the LGP, crossing the Red Sea into the Arabian Peninsula and from there onto the Middle East, Eurasia and Australasia (Templeton 2002; Fernandes et al. 2012).

As the Earth's axis oscillates during sequential precessions of the equinoxes, pointing to different parts of the sky, a bright star of the smallest angular distance from the north celestial pole becomes the North Star. Throughout the Paleolithic a bright North Star would have attracted the attention of the upright standing and walking *H. erectus* and *H. sapiens*, or at least those individuals among the species who were sufficiently attentive. The precession in the northern hemisphere results in Polaris, the 45th brightest star in the nightly sky, to be the Pole Star of today, as it was also c. 26,000 BP, 52,000 BP or 78,000 BP. Even though it never comes closer than 5° off the north celestial pole, the brilliant Vega *was* the northern Pole Star at c. 90,000 BP, 65,000 BP, 40,000 BP and at 14,000 BP. At 40,000 BP and 14,000 BP Vega, observed from the vicinity of the equator and elsewhere north thereof, appeared as a dazzling, unmoving pivot with the nightly sky revolving around it. Intermittently during this period also the stars Deneb and Polaris, as bright North Stars, would attract the attention of the human observer. Even without much observational skill, the bright North Star would come into view of the naked eye as a constant, stationary celestial pivot, with the entire firmament spinning around it throughout each starry night, across the lives and generations of humans. With its exceptional brightness, Vega is the fifth brightest object in the entire nightly sky and is preceded by some 4000 years by the shining Deneb as the North Star. Through each equinoctial cycle, the star Deneb (Alpha Cygni) peaks as the North Star within 5° angular distance, and gradually over the ensuing next four millennia, the brilliant Vega (Alpha Lyrae) peaks also within about the same close angular distance of 5° from the pole (Fig. 2.1).

Starting with *H. erectus* sporadic observations of the nightly sky set in motion migration northward, as if toward the permanent beacon in the sky, the North Star, which during the entirety of a human life had never changed its position in the nightly firmament. Prominently visible incumbent pole stars in the nocturnal sky would have marked direction northward for waves of migrating humans.

Across millennia of the LGP, 115,000–12,000 BP, north of the equator, humans on plateaus and high places where no forests or mountains

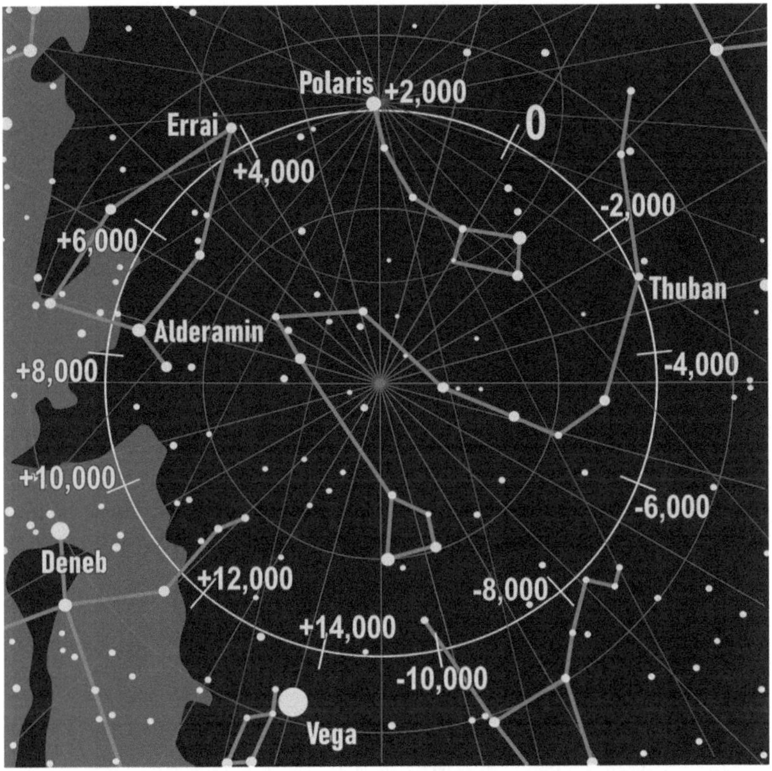

Fig. 2.1 North Stars during current equinoctial cycle, ending 2105 CE. (Drawn by Paul Van Pul)

would hamper their nightly view of the sky, could watch the North Star as the pivot around which the rest of the sky revolves each calm night. When the North Star during a period of the equinoctial cycle was particularly bright, the perceptual effect upon an attentive observer in higher latitudes was immense and would beg a question on the link between the North Star and the circumpolar stars. Anthropomorphic projection of the hunters among the *Homo erectus*, and of the people during the Upper Paleolithic, could lead to only one explanation: North Star controlling the circumpolar stars. Projecting his own physical prowess required in hunting or in the domestication of animals, the hunter would inevitably assign masculinity as the founding feature to the North Star.

Table 2.2 North Stars during current equinoctial cycle, ending 2105 CE

Catalog name	*Namesake*	*Brightness*	*Peak*
Alpha Ursae Minoris	Polaris	1.98	23,600 BCE
Gamma Cephei	Alrai	3.22	21,800 BCE
Alpha Cephei	Alderamin	2.5	18,500 BCE
Alpha Cygni	Deneb	1.25	16,000 BCE
Alpha Lyrae	Vega	0.03	12,000 BCE
Alpha Draconis	Thuban	3.65	2800 BCE
Beta Ursae Minoris	Kochab	2.08	1100 BCE
Alpha Ursae Minoris	Polaris	1.98	2105 CE

Such an explanation could now be extended to the sites of the Venus figurines. Most of the Upper Paleolithic Venuses were carbon-dated to intervals surrounding the peak period of the three bright North Stars, between 40,000 BP and 13,000 BP, with the star Deneb peaking at about 18,000 BP, with Vega peaking at 40,000 BP and 14,000 BP, and with Polaris peaking at about 26,000 BP (Table 2.2). Early north-hemispheric cultures of the Upper Paleolithic that were able to discern the North Star, would be inclined to identify it with a male deity, as an eternal, immutable hero controlling the revolving circumpolar stars in its vicinity and the entire sky. On the steppes of central Siberia, on the plains and high places of east-central Europe and on the coastal areas of western Europe and the British Isles, where topography allowed for a view of the north celestial pole, people could observe each clear night throughout the year, and through their lifetimes, the spectacle of the North Star, fixed as a pivot point in the sky, with all other stars encircling it. Against the masculine sky, represented during day by the sun, and by the North Star during night, the Venus figurines of Eurasia had represented the Earth Mother as the feminine counterpart and a complement.

With gradual warming and receding glaciers, from about 18,000 BP to about 12,000 BP, thus, on shores of seas and lakes facing the north, in highlands and expansive plains north of the equator, an eye-catching observation would have been made time and again during each starry night in the course of the entire year, and by multitudes of humans. Throughout this period of some 6 millennia, the nightly sky along and north of the equator had the North Stars Deneb, at 18,000 BP and then Vega, at 14,000 BP, as salient pivots in the nightly rotation of the firmament. For people north of the equator the presence of the Pole Star, each clear night, as long as the view was not hampered by weather, topography or high vegetation, was a continual observational opportunity throughout the entirety of their lives.

In Eurasia the retreat of glaciers at the close of the Ice Age, past the LGM, allowed nomadic hunters and gatherers to find new sources of fish, game and arable or pasturable land (Richerson et al. 2001: 387). The brilliant North Stars Deneb and Vega would have been instrumental in enticing nomadic human communities to northerly regions of the Earth to seek such new sources of livelihood. The beacon of the North Star along with the retreat of glaciers could help explain the northwardly movement of people both in the west of Europe and in the steppes of eastern Siberia.

The northernmost parts of Eurasia were settled first at northeastern Siberia, which was largely tundra even during the LGM (Barr and Clark 2011). The prehistoric ancestors of the Chukchi at Chukotka, of the Yupik and the Inuit peoples of the northern parts of Siberia and North America, are thought to have arrived at the Bering Sea area sometimes during or after 23,000 BP (Bonatto and Salzano 1997; Raghavan et al. 2015). In contrast, the southern shores of the Baltic Sea were reached only at approximately 15,500 BP when communities of reindeer hunters in Europe, known as the Hamburg culture, had arrived (Riede 2011: 251–264). By about 12,000 BP both northern and southern hemispheres were sparsely inhabited by humans.

At the onset of the Holocene, c. 12,000 BP, Neolithic farmers of the Agricultural Revolution would have recognized the importance of the North Star as the permanent fixture in the sky, always pointing northward to the source of cold winds as well as game and birds heading in the opposite southward direction before the arrival of the winter. Several millennia later, however, at around 4000 BCE, the North Star was no longer a magical brilliant beacon in the nightly sky.

The shift was in the observational change from a brilliant star near the northern celestial pole onto a seeming void in the nightly sky instead of a shining North Star. During the late Neolithic and the Bronze Age, no bright North Star occupied the vicinity of the north celestial pole any longer. Millennia old tradition of stargazing in many cultures had very likely continued into the Neolithic, but with the one major observational deficit in the late Neolithic and the Bronze Age. The presupposition of a feminine *earthbound* goddess myth as a counterpart to the North Star masculine deity would no longer apply with the demise of the bright North Stars. The North Star at 4000 BCE was the faint Alpha Draconis, barely visible to the naked eye during the period when the cup-and-ring stone art was being produced. It is likely that the almost invisible Alpha Draconis, or Thuban as it was known to the Egyptians, substituted for a void in the nightly sky, at the northern celestial pole, round which the firmament was revolving.

The British and European Neolithic cup marks, as well as the central-European round enclosures, are largely on the western edge of the Eurasian geographic belt of the Upper Paleolithic Venus sites (Fig. 2.2). A mutual allegoric association exists between the Venus figurines *and* the later cup-and-ring marks. This association arises as gender is projected as a

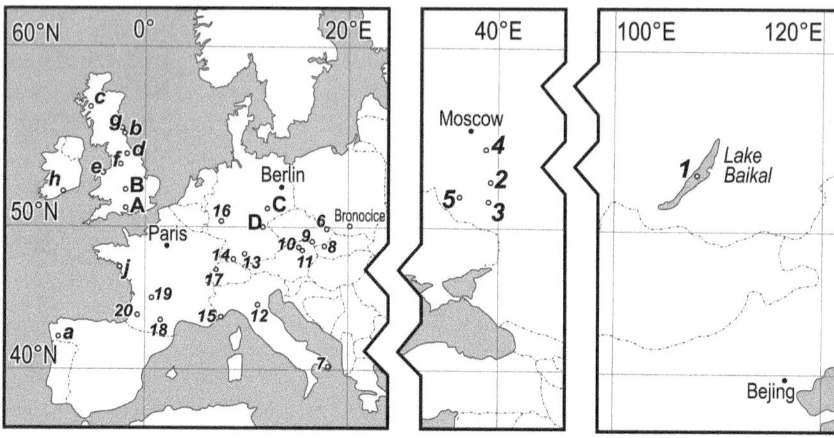

Fig. 2.2 Distribution of Upper Paleolithic Venus sites, Neolithic cup marks and round earthworks through Eurasia. (Drawn by Paul Van Pul)

Upper Paleolithic Venus sites, 1–20, see Table 2.1

Major cup-and ring-sites (lower case italic)
a Ponte Caldelas 42° N 9° W
b Alnwick 55° N 2° W
c Kilmartin Glen 56°N 5°W
d Ilkley Moor 54°N 2°W
e Bachwen 53°N 4°W
f Anderton 53°N 2°W
g Weetwood Moor 56°N 2°W
h Derrynablaha 52° N 10° W
j Arzon 48°N 3°W

Main earthwork or lithic round enclosures (Capital case)
A STONEHENGE 51°N 2°W
B AVEBURY 52°N 2°W
C GOSECK 51°N 12°E
D BYLANY 50°N 15° E

cosmogonic attribute upon the environment, this being the case of the Paleolithic Venuses, during periods of bright North Stars standing for the masculine in the sky, as well as in the case of the cup-and-ring marks, when celestial void may have given a reason to think of the nightly sky as feminine. The proposition by Castleden (1987: 74) suggesting that the cup-and-ring marks are representations of circumpolar stars' diurnal revolution about the celestial pole, further reinforces the feminine interpretation of the cup marks.

2.3 *Axis Mundi* as Architectural Archetype

Cup-and-ring marks are found both throughout the northern and southern hemispheres. Their unusual concentration, in the hundreds, on the British Isles and on the Atlantic coast of Europe suggests, that the early interest in the skies, albeit universal across the world, may have been anchored in the continual visibility of the celestial pole. The pattern of geographic latitude corresponding to optimal locations for viewing the north celestial pole points to anthropomorphic projection as the association between the cup-mark sites and the Venus finds in the Atlantic region of Europe. The Paleolithic stone art of the Venuses on the one hand, and the late Neolithic cup marks representing the diurnal appearance of circumpolar stars, on the other hand, further implicate geographic latitude in the emergence of both kinds of stone art. At the absence of clearly visible North Star during the late Neolithic the stamp of celestial deity controlling the firmament was the void in place of the north celestial pole. In a possible shift in the primordial sky allegory, cup-and-ring carvings, therefore, likely represent revolving circumpolar stars and the setting and rising of all other stars, as a feminine feature of the universe.

Consistent with the celestial void represented by the almost invisible North Star during the late Neolithic, c. 5000–3000 BCE, many cup-and-ring stone carvings, in addition to concentric circles, contain a centermost depression, a designed void, suggesting a feminine feature as a pivot around which the firmament revolves. Material manifestation of the feminine as a sunken interstice in the center of the cup-and-ring mark corresponds to the void at north celestial pole of the late Neolithic.

Common to many of the hundreds of sites of the concentric rings is their positioning near high places or coastlines with exposure northward, unrestricted by landscape or woods, in a geographic latitude comfortable enough for continuous view of the north celestial pole and circumpolar stars during night. In the geographic vicinity of the cup-and-ring marks across England and Scotland circular or semi-circular concentric stone-work had been sometimes built between the late Neolithic and the Bronze Age, often at or beneath hilltops. Believed to be associated with a renewal sky ritual, “rock surfaces decorated with extensive areas of cup-and-ring marks are among the most enigmatic aspects of Bronze Age religion, [such as] sun worship, or indeed associated with other forms of astronomical observation” (Ritchie and Ritchie 1991: 72).

Astronomical observation, possibly for calendar determination, as related also to renewal rituals have been the explanation put forward for the European round enclosures (Biehl 2010; Pasztor et al. 2015). The European round enclosures, uncovered in their dozens only over the last 150 years, mostly through aerial photography, are dated to about 5000–3400 BCE (Milisauskas 2011). The British cup-and-ring marks are believed to have been produced during and after 4000 BCE (Sharpe et al. 2008: 5), while various circular earthworks on the British Isles were built in the period 3000–1500 BCE (Cunliffe 2013: 181–183).

Andrew Jones pointed out that sedentary communities of the Neolithic had a ceremonial space dedicated to possible fertility or other renewal rites within their near proximity (Jones 2011: 89). The design of such primordial civic space as winter-solstice renewal ritual sites would have been most likely inspired by the cyclical renewal observed in the sky, Stonehenge and some of the European round enclosure being examples (Ruggles 1997; Biehl 2012). This is also consistent with other accounts made recently, pointing to ‘psychological projection’ as a link between the nightly sky and prehistoric rock-art (Christian 2014: 26).

It was the altered cosmogonic belief arising from the north polar celestial void that gave rise to cup-and-ring marks with a central depression, possibly also to lithic or ditched enclosures with a central open space, and the multiple invention of the wheel-and-axle between 4000 BCE and 2000 BCE in central Europe, in Mesopotamia and the Caucasus.

The Neolithic cup-and-ring stone art of concentric circles has often a carved linear channel penetrating from outside the perimeter of the mark into its central depression. The linear channel running between the perimeter of the ring and its center appears to substitute for a vertical column, an axle as it were, pointing toward the northern celestial pole. Such column, or the groove substituting for it, represents the axle of the world, the *axis mundi*. The linear groove could be interpreted as conferring either a masculine feature upon the cup-and-ring mark or a feminine feature of an umbilical cord connecting the sky with the Earth.

Similar to the carved straight grooves leading into the center of cup marks, some of the Neolithic round enclosures had access channels lined with standing stones that formed "directing axes of movement into and out of the valley and funneling observers towards significant burial or ceremonial monuments" (Jones 2011: 89). Consistency between the layout of the megaliths and the design of the cup-and-ring marks reinforces the presumption of *axis mundi* as an overriding and universal notion that came to be occasionally expressed by grooves in cup marks and by channeling avenues in round enclosures.

Still today numerous places in the northern hemisphere carry names alluding to the Earth's navel (Leeming 2005: 404). Invoking North American indigenous beliefs the historian of religion, Mircea Eliade, had pointed to the *Axis mundi* as something of a primordial mental template incorporating different realms of the cosmos (Eliade 1959: 36–47). This is echoed in the words of Christian Norberg-Schulz who views the *Axis mundi* as an archetypal myth at the founding of architecture:

> The vertical, therefore, has always been considered the sacred dimension of space. It represents a 'path' towards a reality which may be 'higher' or 'lower' than daily life, a reality which conquers gravity, that is, earthly existence, or succumbs to it. The axis mundi is thus more than the centre of the world, it represents a connection between [...] cosmic realms, and it is only at the central axis that a breakthrough from one realm to another can occur. (Norberg-Schulz 1971: 21)

It is also noteworthy that the very first extant evidence of a wheel is in the bird's-eye view of a wagon incised on a vase from southern Poland. Carbon-dated to c. 3500 BCE the vase comes from Bronocice, at about 50° N and 20° E, within the geographic corridor of the earlier, Paleolithic Venus sites, and also within the central-European region abundant with ditched round enclosures. The cup-and-ring marks from the Atlantic coast and the wheel-and-axle from Bronocice, very likely expressed a new outlook of the north celestial pole, an alternative anthropomorphic explanation of the universe, different from the one associated with the Upper Paleolithic Venuses, and one that also likely inspired the multiple inventions of the wheel.

It is in the latter part of the Neolithic period that the cup-and-ring stone art appears. The cup-and-ring carvings and the plethora of Neolithic round enclosures throughout central and western Europe suggest consistent observational effort to witness the revolving nightly firmament. Such opportunity, clearly, was afforded not only to the inhabitants of Eurasia but also to Paleo-Indians of the Great Plains as North American medicine wheels suggest.

Medicine wheels, also known as sacred hoops, have been built across the Great Plains in their hundreds by ancestral First Nations people since about 3500 BCE (Bryan 2015: 86–123). Much as in the case of ceremonial staging at concentric stone rings on the British Isles and at the round enclosures of central Europe, medicine wheels too seem to have been associated with the scenography for dramatic stagecraft, fulfilling a similar proto-civic function as their European counterparts.

At Stonehenge, as well as at some other lithic sites such as the nearby Avebury, a straight path, an 'avenue' lined at the time with standing stones, leads into the circular enclosure. Similar access path pattern can be discerned also in some North American medicine wheels (Comer 1996: 279). A cairn at the center of the medicine wheel might signify the world's navel (Eaton 2014: 52), while a straight avenue leading to a medicine wheel possibly represents, much as in similar such avenues at Avebury, Stonehenge or in the European round enclosures, the world's axle. Similar to the European round enclosures, Archaic medicine wheels are believed linked to Paleo-Indian celestial observations while serving also as sites for healing and dais for ritual-ceremonial performances.

Paleo-Indians likely did not possess a significant tradition grounded in mythology of the North Star. Early human migrations east and northward, had reached Beringia, an enormous land-and-ice mass then connecting northeastern Siberia and northwest Alaska, sometime after 20,000 BP. Raghavan et al. (2015) point to a period of isolation of 8000 years, or more, of the Siberian migrants into America trapped at the Beringia land bridge. Raghavan's Beringian Incubation Model explains why people who crossed the Beringia into North America, those whose descendants came to be called the Paleo-Indians, may never have produced Venus figurines: during much of the initial peopling of North America, there were no bright North Stars to signify masculinity in the nightly sky. Further, at approximately 70° northern latitude any North Star that might have been easily visible to the naked eye, from what is today the Bering Strait, would be almost directly overhead, providing no mythical direction to wandering bands of humans. In circumpolar regions, it was the direction in which hunters followed big game rather than a stationary beacon near the center of the nightly sky that had determined the movement of people. In much of Eurasia, however, the northern celestial pole was at the founding of civilizations.

From linear grooves in cup-marks to straight access routes of circular stonework, the archetypal template of the *Axis mundi* came to be carved in the minds of humans into the Bronze and Iron Ages, heralding the layout of planned streetscapes, as well as a common theme in some early religions:

> All sky gods have certain high places set apart for their worship. The symbolic and religious significance of mountains is endless. Mountains are often looked on as the place where sky and earth meet, a "central point," therefore, the point through which the Axis Mundi goes, a region impregnated with the sacred, a spot where one can pass from one cosmic zone to another. So, in Mesopotamian belief, "the Mountain of the Lands " unites earth and heaven, and in Indian mythology Mount Meru rises up in the centre of the world; above it the Pole Star sends forth its light. The Uralo-Altaic peoples also have a central mountain, Sumbur, Sumur or Senieru, above which hangs the Pole Star. (Eliade 1958: 100)

Christian Norberg-Schulz extends Eliade's sagacity further pointing to the inherent bond between *Axis mundi* and the geometric notion of the center. Such linkage goes beyond religious belief, as it identifies an essential mode of human consciousness in the perception of space. To Norberg-Schulz the linkage ingrained in the bond between the concept of the center and the myth of the *Axis mundi* is at the very foundation of the built environment:

> [T]he gable post in early wooden construction supported the ridge beam at each end of a building. Its primary role led to its Old High German name *Irminsul* (universal column), that which supports all things. Analogously, the ridge beam was the 'celestial axis.' (Norberg-Schulz 1996: 52)

Aligned with the north-south axis and intersecting at right angles were the straight, long streets of Harappa and Mohenjo-Daro, two prominent cities of the Indus Valley Civilization, about 2600 BCE. Attesting to the significance of the Harappan Civilization's aligned streets is evidence of their paving, likely the earliest extant, possibly pointing to occasional use of the streets also as processional avenues (Dani and Thapar 1996). Attributed to Hippodamus of Miletus (498–408 BCE), the 'first city planner' according to Aristotle, is northern alignment of gridiron street layout in the Greek colony Thurii, founded 452 BCE on the Italian peninsula, and in the archaic Ionian city of Priene, rebuilt around the same time. All major Roman colonies were also planned on a grid pattern, aligned with the *cardo maximus*, the main avenue running in the north-south direction.

2.4 *Ouroboros* and the Rise of Public Place

As opposed to the gridiron pattern stands the round place. It is ultimately through the notion of the center implicit in the myth of the *Axis mundi* that Norberg-Schulz finds the communal space, as a round *place*, an enclosure open to the sky:

> For its definition, therefore, the place needs a pronounced limit or border. The place is experienced as an "inside" in contrast to the surrounding "outside." [...] A place, therefore, is basically "round." [...] (Norberg-Schulz 1971: 20)

As far as deliberate urban design is concerned, only very few cities in antiquity were planned on a radial or some other circular pattern. Whereas to many a pundit, a grid pattern of a delineated space may seem to endorse a mechanistic proclivity, the circular place, from the Neolithic round enclosure to the Greek and Roman theater, has been associated more with a dramaturgical performance and ritual:

> The enclosure has strong social implications. Basically, it expresses a coming together, the forming of a ring for a common purpose. Most cultures have such [open-air] enclosures where rituals or theatrical performances take place. (Goodsell 1988: 36)

In at least some of the Neolithic roundels, public fertility rituals took place (North 1996: 358; Milisauskas 2011), the pregnant female manifesting the mystery of renewal. In their circular layout the round earthworks, along also with the cup-and-ring stone art, are archetypal projections of rebirth and eternal renewal, heralding the *Ouroboros* cyclical renewal myth later in the Bronze Age. The winter solstice and the periodic fertility rituals at staging sites at the round enclosures could thus be also seen as presaging the *hieros gamos* enactment.

Benjamin's anthology of urban observations, *One Way Street*, written in 1928, has its last section 'To the Planetarium,' addressing the relation of archaic people with their universe. In the midst of urban essays, this might seem like an aberration. But the connection of the essay to the overall urban context of the anthology is Benjamin's lament over the loss to urban civilizations through the many millennia since the Neolithic:

> Nothing distinguishes the ancient from the modern man so much as the former's absorption in a cosmic experience scarcely known to later periods. […] All the same, the exclusive emphasis on an optical connection to the universe, to which astronomy very quickly led, contained a portent of what was to come. The ancients' intercourse with the cosmos had been different: the ecstatic trance. (Benjamin 1928, in Ferris 2008: 75–77)

The monumental stonework at Stonehenge and Avebury exemplify Benjamin's statement as sites designated for burials, as well as the staging grounds of fertility rites (Tilley 2010: 285–292). The Wiltshire stonework had served as astronomical observatories to mark the winter solstice, similar to even older ditched round enclosures in central Europe. The view that these were designed public places of community encounter may be seen reinforced by evidence of carefully aligned access routes that at least in some cases, the Goseck Circle in Germany being one, were in the northerly direction, tracing the *Axis mundi*. The Atlantic cup marks and the European round enclosures were forerunners of the renewal myth of *Ouroboros* and with it also the sexual rituals of *hieros gamos* later in the Bronze and Iron Age (Fig. 2.3).

Fig. 2.3 Goseck Circle, reconstruction with access route tracing the *Axis mundi*, and sunrise and sunset gates for winter solstice, near Leipzig, Germany. (Credit: Ralph Beutnagel)

The emergence of the myth of the Earth Mother, too, ought to be seen concordant with a celestial masculine myth focused on the North Star precisely within the emergence of the Eternal renewal archetype. The earthbound renewal myth of a life-giving female deity and the sky-bound masculine myth of a deified hero, provider and protector, came to be complementing one another. Projection of the most tangible of human features, that of gender, upon the perceptible environment had yielded the anthropomorphic opposites of the celestial father deity against the terrestrial mother goddess.

The archetypal stature of the Eternal Return is strikingly evident through the struggle that Benjamin wages against it. Benjamin repeatedly rejects it, only to return and argue against it, time and again in Convolute D (*Boredom, eternal return*, pp. 101–119) of the *Arcades Project*:

> The belief in progress – in an infinite perfectibility understood as an infinite ethical task – and the representation of eternal return are complementary. They are the indissoluble antinomies of the face of which the dialectical conception of historical time must be developed. In this conception, the idea of eternal return appears precisely as that "shallow rationalism" which the belief in progress is accused of being, while faith in progress seems no less to belong to the mythic mode of thought than does the idea of eternal return. (Benjamin 1933/1999: 119)

Most of the Venus figurines known today were unearthed only after Benjamin's untimely death in 1940. One may only ponder where Benjamin's captivating interest in the cosmos and in the dawn of civilization, would lead him had the archeological discoveries of the Venuses occurred only a few decades earlier.

Benjamin was also unaware of the European round enclosure, most of which too were uncovered only in the latter part of the twentieth century. Many prehistoric circular enclosures found in the northern hemisphere were often associated with renewal rituals of burial and fertility and sometimes astronomically aligned to the annual solstices (Mann 2011: 143–146). The prehistoric humans' view of the cyclicality of observed celestial patterns likely drew on the projection of both masculine and feminine features as acumen in astronomical discernment of the sky.

With the demise of Vega as the North Star, a prominent masculine sign in the nightly sky was diminished, and there also would be lesser mythical drive to produce a feminine symbolism as a counterpart. All the same, however, outlook on the early observations of cyclicality and gender projection had remained universal across the northern hemisphere. Eternal renewal was behind the archaic yin-yang dichotomy of the Chinese which, finding its way to ancient Egypt, transmuted into the myth of the *Ouroboros*, a snake in a circular position devouring its own tale:

> […] yin and yang represent creation, and the ouroboros constantly recreates itself. Yin is symbolic of earth, female, dark, and negative, while yang corresponds to heaven, male, light, active and positive. To create a balance, each half needs to feed off the other. (Cox 2009: 152)

Gender imprints had been projected onto archaic myths, as well as upon early built environments. Archaic medicine wheels through the Great Plains of North America and Neolithic round enclosures through Europe allude not only to the cyclicality of the Eternal Return but also to the *Axis mundi*, the world axle, sometimes as a masculine feature, other times as feminine representation of the world's umbilical cord. Paleolithic Venus figurines and Neolithic cup-and-ring marks further reinforce the view of continual gender projection upon material culture. The subsequent appearance of Neolithic round enclosures in Eurasia and of archaic medicine wheels in the Great Plains, strengthen the presumption of prehistoric ritual sites evolving into primordial community cultic places where yearly renewal-rebirth-healing, or eternal return, were marked. The gender interpretations of the linear carving in cup marks suggest correspondence of the *Axis mundi* with the divine union myth, in a setting also presaging the *hieros gamos* ritual during the Bronze Age and later.

Erich Neumann, "Jung's anointed intellectual heir" (Paglia 2006), highlights this very same link in regard to the archetypal notion of the Eternal Return as it had manifested itself in myth during the Bronze and Iron Age:

> The *ouroboros* appears as the round "container", i.e. the maternal womb, but also as the union of masculine and feminine opposites, the World Parents joined in perpetual cohabitation. (Neumann 1949/1954: 13)

Neumann had directed much of his considerations of the evolution of consciousness to mythologies of the goddess identifying the Paleolithic Venus figurines as representing the primordial origin of later varieties of the Great Mother. Yet far from claiming a Paleolithic matriarchal religion, indications are that Neumann in fact contemplated to identify also masculine features in early cult. It was only due to his early death in 1960 that he was not given the chance to pursue the task he set out to himself to accomplish. To Neumann, the Great Mother archetype was eventually conjoined by the mythic Hero archetype, to yield a transcendent gender dichotomy.

On October 7, 1950, Neumann held a lecture at the Psychological Club of Zurich titled 'Zur Psychologie des Weiblichen im Patriarchat' ('Toward a Psychology of the Feminine in Patriarchy'), which he repeated in Basel and Tel Aviv (Liebscher 2015: 198). It is, therefore, very likely that Neumann saw his volume, *The Great Mother*, as only one part of a more comprehensive concept which would have involved a patriarchal complement.

Drawing on the yin and yang dichotomy in Chinese myth Neumann points to gender opposites as parallel to other binary appearances, such as the diurnal cycle, day and night. Citing Ernst Cassirer in *The Origins and History of Consciousness*, Neumann contends that the built environment is the projection of opposites:

> World-building, city-building [...] are all reflection of the original mythology of space, which, beginning with the opposition between light and darkness, classifies and arranges the world in a continuous series of opposites. [...]The sacred world order and the sacred space – precinct or sanctuary – were 'oriented' by this opposition. (Neumann 1949/1954, 167–8)

To Jung, to the antiquarians Marija Gimbutas and Karol Kerenyi and later to Erich Neumann, the Paleolithic Venus figurines are evidence to the Earth Mother stature as a dominant myth (Jung and Kerenyi 1951/1977, CW 9i, ¶ 312). The Earth Mother was the primeval goddess, her myth leading, through the *hieros gamos* rituals, later in western religion into transformation to the Christian Mother of God (Dourley 1984: 51–70).

2.5 Eternal Return and *Axis Mundi* as Gendered Myths

In Greek myth of the Iron Age (c. 1200 BCE), Gaia, the Great Mother and the primordial goddess of the Earth is in a marital union with Uranus, the god of sky. The sky has been the source of many creation allegories, much as the earth, too, has been a primordial paradigm of various cosmogony myths. A Greek creation myth that appears to link the Earth with the sky is the account of Pandora, the first woman. Pandora was molded from *earth* but was created on order of Zeus, the god of the *sky* and thunder, the grandson of Uranus. In *Works and Days* Hesiod, the Greek poet of the seventh century BCE, relates the earliest literary version of Pandora, as a female amalgam of seductive gifts of gods that were actual afflictions, meted on mankind in punishment for accepting stolen fire by Prometheus the Titan (*Works and Days* 60–105). Pandora exerts her power through a jar in what appears as feminine symbolism. Out of sheer curiosity, rather than malice, Pandora is said to have opened a jar that contained toil, sickness, disease and all other evils of mankind, releasing them all but one: hope. As a first mortal woman, she thus appears to attain supremacy through her femininity, somewhat akin to a priestess.

It was the priestess, or a cult associated with one, that appears to have inspired the Minoan civilization on the Mediterranean island of Crete during the Bronze Age, c. 2700–1550 BCE. The Minoan priestess ritual had been performed in sanctuaries at caves or at designed open spaces atop mountains and archeological finds on Crete point to a prominently gynocentric Minoan culture and religion. In classical Greece, a millennium later shrines such as Delphi, have followed in the Minoan female priestess tradition and also the cult of the bisexual god Dionysus representing femininity in later classical Greece has been traced to Minoan Crete (Kerenyi 1976: 63–128).

An important feature of many Minoan sanctuaries on Crete, such as the open-air peak sanctuary on Mt. Iuktas and the adjacent palace at Knossos, is their alignment with adjacent palaces on a north-south axis (Marketou 1988: 28–31; Karetsou 2010). Another sanctuary, the sacred cave of Kamares at Mt. Ida, similarly, has a north-south alignment with the palace at Phaistos. Consistent with the numerous Minoan peak sanctuaries, a north-south orientation can also be found in many Egyptian pyramids, burial monuments constructed around the same time as or a few hundred years earlier than the Minoan ritual spaces.

At c. 2700 BCE a north-south orientation could be attained most expediently by alignment with the star Alpha Draconis, or Thuban to the Egyptians, as the North Star. Alpha Draconis is a faint star of brightness 3.7, and even if it was brighter c. 2700 BCE than it is today, it still would have required an observational skill to be identified as the North Star. There are no bright stars in the sky vicinity of Alpha Draconis. To the Egyptians, the dark area in the nightly sky in the vicinity of Thuban was considered void around which the rest of the universe revolves. After burial, Pharao's soul was believed to ascend to the abode of gods, and the pyramids were designed specifically for the magical purpose of launching the soul from its burial place through this celestial void.

There are no extant written records by the Minoans but it is fair to presume that, due to lack of any clearly visible stars about the celestial North Pole at the time, common Minoans might have considered the celestial north, a dark area around which the rest of the celestial sphere turns, as a great void. The configuration of Minoan ritual places, as open area peak sanctuaries pointing to the north could be explained, thus, as an expedience to the priestesses linking the void in the sky with the female on earth.

Some of the Minoan culture and religion was likely influenced by Egypt, from the south, and by Mesopotamia, from the east. In Mesopotamian myths of the Bronze Age (c. 3000 BCE), Kishar, the Earth Mother, is the wife of Anshar, the archaic god of heavens. Further east, in the Hindu religion, the union of male deity Shiva with the female goddess Shakti represents the merger of a masculine divinity, limitless and unchanging, with cosmic femininity, a deity manifest by uncontrollable energy. Carl Jung had pointed out that in contrast to the union of deities and the unity of the world, multiplicity of the world arises in Hindu belief from splitting of the divine cosmic union of femininity and masculinity (Jung 1950/1977 CW 9i, ¶ 632). In some early societies, Jung explains further, the matrimony of gods and goddesses had signified primeval religious proceedings toward union of opposites (Jung 1951/1978a CW 9ii, ¶ 425). Mimicking this union of gods and goddesses across archaic cultures of the northern hemisphere was the *hieros gamos* sexual ritual (Jung 1952/1975 CW 11, ¶ 748).

Erich Neumann had pointed out the link between the Great Mother religion and the *Ouroboros* cyclical renewal myth on the heels of which also the *hieros gamos* came (Neumann 1949/1954: 5–38). The earliest extant specimen of the *Ouroboros* is a round image of serpent deities in world's periodic renewal, carved as part of an Egyptian coffin text, which millennia later in medieval imaging transformed into the prevailing and by now-common image of a snake or a dragon swallowing their own tails.

A major perceptual determinant implicated in the emergence of the *Ouroboros* and the ensuing myths had been the circumpolar stars in their diurnal revolution round a bright North Star, or in the absence of one, round a perceived dark void of the north celestial pole. The advent of the *Ouroboros* in cup marks and Neolithic round enclosures is by necessity complemented by the *Axis mundi*, the world axle. The one cannot exist without the other: the Eternal Return implies a cyclical, circular motion, and thus also an axle or axis, while an axle without a pendant attachment, similarly, is impossible to fathom.

The suggestion that the Eternal Return is an archetypal notion, a mental template ingrained in the minds of humans arises from a comment made by C.G. Jung in his Seminar on Nietzsche's *Thus Spake Zarathustra*. In commenting on Nietzsche's adherence to the Myth of the Eternal Return, Jung stated that the ring of eternal recurrence is an individuation symbol, apparently taken as an archetypal template (Jung and Jarrett 1939/1989: 1044). A corollary ought to be made to the effect that *Axis mundi* is also an archetypal notion, and the two complement each other. In the built environment of the early planned settlements of the Bronze Age, *Axis mundi* found its own manifestation in the aligned street and the straight avenue, much in the vein of the access avenues that had led to round enclosures, to the megaliths at Stonehenge and Avebury or to archaic medicine wheels in the Great Plains. Imprinted in the urban plans of antiquity as a universal archetypal template, is a perceived bond between the perfection of the universe and the wisdom in the plan of an ideal city,

> "celestial axis" around which the world is ordered, a conception that corresponds to the Roman cardo that runs from the North Star to the Southern sun. (Norberg-Schulz 1996: 52)

Bibliography

Akkerman, Abraham. 2016. *Phenomenology of the Winter-City: Myth in the Rise and Decline of Built Environments*. New York/Dordrecht/London: Springer.

Aveni, Anthony F. 2001. *Skywatchers: A Revised and Updated Version of Skywatchers of Ancient Mexico*. Austin: University of Texas Press.

Barr, Iestyn D., and Chris D. Clark. 2011. Glaciers and Climate in Pacific Far NE Russia During the Last Glacial Maximum. *Journal of Quaternary Science* 26 (2): 227–237.

Benjamin, Walter. 1933/1999. *The Arcades Project*. Trans. H. Eiland and K. McLaughlin. Cambridge, MA: Belknap.

Biehl, Peter F. 2010. Measuring Time in the European Neolithic? The Function and Meaning of Central European Circular Enclosures. In *The Archaeology of Measurement: Comprehending Heaven, Earth and Time in Ancient Societies*, ed. Iain Morley and Colin Renfrew, 229–244. Cambridge: Cambridge University Press.

———. 2012. Meaning and Function of Enclosed Places in the European Neolithic: A Contextual Approach to Cult and Religion. *Archaeological Papers of the American Anthropological Association* 21 (1): 130–146.

Bonatto, Sandro L., and Francisco M. Salzano. 1997. A Single and Early Migration for the Peopling of the Americas Supported by Mitochondrial DNA Sequence Data. *Proceedings of the National Academy of Sciences* 94 (5): 1866–1871.

Bryan, Liz. 2015. *Stone by Stone: Exploring Ancient Sites on the Canadian Plains.* Victoria/Calgary: Heritage House.

Castleden, Rodney. 1987. *The Stonehenge People: An Exploration of Life in Neolithic Britain 4700-2000 BC.* London/New York: Methuen.

Christian, Thomas. 2014. The Night Sky, Psychological Projection, and Ancient Rock Art: A View into the Human Psyche. *Jung Journal: Culture & Psyche* 8 (4): 26–33.

Comer, Douglas C. 1996. *Ritual Ground: Bent's Old Fort, World Formation, and the Annexation of the Southwest.* Berkeley/Los Angeles: University of California Press.

Cox, Simon. 2009. *Decoding the Lost Symbol: The Unauthorized Expert Guide to the Facts Behind the Fiction.* New York/London: Simon and Schuster.

Cunliffe, Barry. 2013. *Britain Begins.* Oxford: Oxford University Press.

Dani, Ahmad Hasan, and B.K. Thapar. 1996. The Indus Civilization. In *UNESCO History of Civilizations of Central Asia: The Dawn of Civilization, Vol II Earliest Times to 700 B.C.*, ed. A.H. Dani, 271–307. Paris/New York: UNESCO.

Dorsey, George Amos. 1997. *The Pawnee Mythology.* Lincoln: First Bison Books, University of Nebraska Press.

Dourley, John P. 1984. *The Illness That We Are: A Jungian Critique of Christianity.* Toronto: Inner City Books.

Dunne, Pete. 2011. *Arctic Autumn: A Journey to Season's Edge.* New York: Houghton Mifflin Harcourt.

Eaton, Evelyn. 2014. *The Shaman and the Medicine Wheel.* Wheaton: Theosophical Publishing House.

Eliade, Mircea. 1958. *Patterns in Comparative Religion.* Trans. Rosemary Sheed. London: Sheed and Ward.

———. 1959. *The Sacred and the Profane: The Nature of Religion.* New York: Harcourt, Brace & World.

Endicott, Phillip, S.Y.W. Ho, and C. Stringer. 2010. Using Genetic Evidence to Evaluate Four Palaeoanthropological Hypotheses for the Timing of Neanderthal and Modern Human Origins. *Journal of Human Evolution* 59 (1): 87–95.

Fernandes, Verónica, Farida Alshamali, Marco Alves, et al. 2012. The Arabian Cradle: Mitochondrial Relicts of the First Steps Along the Southern Route Out of Africa. *American Journal of Human Genetics* 90 (2): 347–355.

Ferris, D.S. 2008. *The Cambridge Introduction to Walter Benjamin.* Cambridge/New York: Cambridge University Press.

Goodsell, Charles T. 1988. *The Social Meaning of Civic Space: Studying Political Authority Through Architecture.* Lawrence: University Press of Kansas.

Griffin-Pierce, Trudy. 1992. *Earth Is My Mother, Sky Is My Father: Space, Time, and Astronomy in Navajo Sandpainting.* Albuquerque: University of New Mexico Press.

Hammer, Michael F., August E. Woerner, Fernando L. Mendez, Joseph C. Watkins, and Jeffrey D. Wall. 2011. Genetic Evidence for Archaic Admixture in Africa. *Proceedings of the National Academy of Sciences of the United States of America* 108 (37): 15123–15128.

Holt, Brigitte M. 2015. Chapter 13: Anatomically Modern Homo sapiens. In *Basics in Human Evolution*, ed. Michael P. Muehlenbein, 177–192. London: Elsevier.

Jones, Andrew. 2011. Biographies in Stone: Place, Memory and the Prehistory of Sculpture. In *Sculpture and Archaeology*, ed. Paul Bonaventura and Andrew Jones, 83–96. Surrey: Ashgate.

Jung, C.G. 1950/1977. Concerning Mandala Symbolism. *The Archetypes and the Collective Unconscious.* CW 9i.

———. 1952/1975. Answer to Job. Lectori Benevolo. *Psychology and Religion: West and East.* CW 11.

———. 1954/1975. On the Nature of the Psyche. *The Structure and Dynamics of the Psyche.* CW 8.

———. 1955/1970. The Conjunction. The Third Stage: The *unus mundus. Mysterium Coniunctionis.* CW 14.

———. 1951/1978a. Conclusion. *Aion: Researches into the Phenomenology of the Self.* CW 9ii.

———. 1951/1978b. The Sign of the Fishes. *Aion: Researches into the Phenomenology of the Self.* CW 9ii.

Jung, C.G., and James J. Jarrett, eds. 1939/1989. *Nietzsche's Zarathustra: Notes of the Seminar Given in 1934–1939,* vol. 2, 1044. Princeton University Press.

Jung, C.G., and Karl Kerenyi. 1951/1977. The Psychological Aspects of the Kore. *The Archetypes and the Collective Unconscious.* CW 9i.

Karetsou, Alexandra. 2010. Popular Cult and Power: The Function of the Juktas Peak Sanctuary in Protopalatial and Neopalatial Crete, Lecture, 26 February 2010. Dublin: Trinity College.

Kerenyi, Carl. 1976. *Dionysos: Archetypal Image of Indestructible Life.* Princeton: Princeton University Press.

Leeming, David Adams. 2005. *Oxford Companion to World Mythology.* Oxford: Oxford University Press.

———. 2010. *Creation Myths of the World: Part III.* Santa Barbara: ABC-CLIO.

Li, Ying-Chang, and Eva Wong. 1994. *Lao-Tzu's Treatise on the Response of the Tao.* Walnut Creek: Alta Mira Press.

Liebscher, Martin. 2015. *Analytical Psychology in Exile: The Correspondence of C. G. Jung and Erich Neumann.* Princeton/Oxford: Princeton University Press.

MacDonald, John. 1998. *The Arctic Sky: Inuit Astronomy, Star Lore, and Legend.* Toronto: Royal Ontario Museum/Nunavut Research Institute.

Mann, Nicholas. 2011. *Avebury Cosmos: The Neolithic World of Avebury Henge, Silbury Hill, West Kennet Long Barrow, the Sanctuary & the Longstones Cove.* Alresford: John Hunt Publishing.

Margold, Martin, John D. Jansen, Artem L. Gurinov, Alexandru T. Codilean, David Fink, Frank Preusser, Natalya V. Reznichenko, and Charles Mifsud. 2016. Extensive Glaciation in Transbaikalia, Siberia, at the Last Glacial Maximum. *Quaternary Science Reviews* 132: 161–174.

Marketou, T. 1988. New Evidence on the Topography and Site History of Prehistoric Ialysos. In *Archaeology in the Dodecanese*, ed. Soren Dietz and Ioannis Papachristodoulou, 28–31. Copenhagen: The National Museum of Denmark, Department of Near Eastern and Classical Antiquities.

Milisauskas, Sarunas. 2011. Early Neolithic, the First Farmers in Europe, 7000–5500/500 BC. In *European Prehistory: A Survey*, ed. S. Milisauskas, 153–206. New York/Dordrecht: Springer.

Neumann, Erich. 1949/1954. *The Origins and History of Consciousness*. With a Foreword by C. G. Jung. London: Routledge & Kegan Paul.

Nicolson, Iain. 1999. *Unfolding Our Universe*. Cambridge: Cambridge University Press.

Niessen, Frank, Jong Kuk Hong, Anne Hegewald, Jens Matthiessen, Rüdiger Stein, Hyoungjun Kim, Sookwan Kim, Laura Jensen, Wilfried Jokat, Seung-Il Nam, and Sung-Ho Kang. 2013. Repeated Pleistocene Glaciation of the East Siberian Continental Margin. *Nature Geoscience* 6: 842–846.

Norberg-Schulz, Christian. 1971. *Existence, Space & Architecture*. New York: Praeger Publishers.

———. 1996. *Nightlands: Nordic Building*. Cambridge, MA: MIT Press.

North, John David. 1996. *Stonehenge: A New Interpretation of Prehistoric Man and the Cosmos*. New York: Free Press.

Paglia, Camille. 2006. Erich Neumann: Theorist of the Great Mother. *Arion* 13 (3): 1–14.

Pasztor, Emılia, Judit P. Barna, and Georg Zotti. 2015. Neolithic Circular Ditch Systems ("Rondels") in Central Europe. In *Handbook of Archaeoastronomy and Ethnoastronomy*, ed. C.L.N. Ruggles, 1317–1326. New York: Springer.

Raghavan, Maanasa, et al. 2015. Genomic Evidence for the Pleistocene and Recent Population History of Native Americans. *Science* 349 (6250): aab3884.

Richerson, Peter J., Robert Boyd, and Robert L. Bettinger. 2001. Was Agriculture Impossible During the Pleistocene But Mandatory During the Holocene? A Climate Change Hypothesis. *American Antiquity* 66 (3): 387–341.

Riede, Felix. 2011. Steps Towards Operationalising an Evolutionary Archaeological Definition of Culture. In *Investigating Archaeological Cultures: Material Culture, Variability and Transmission*, ed. Benjamin W. Roberts and Marc Vander Linden, 245–270. New York/Heidelberg: Springer.

Ritchie, James N.G., and Anna Ritchie. 1991. *Scotland: Archaeology and Early History*. Edinburgh: Edinburgh University Press.

Ruggles, Clive. 1997. Astronomy and Stonehenge. In *Science and Stonehenge*, Proceedings of the British Academy 92, ed. Barry Cunliffe and Colin Renfrew, 203–229. Oxford: Oxford University Press.

Sharpe, Kate, Tertia Barnett, et al. 2008. *England's Rock Art. The Prehistoric Rock Art of England*. Morpeth, Northumberland: English Heritage & Northumberland County Council.

Templeton, Alan. 2002. Out of Africa Again and Again. *Nature* 416: 45–51.

Tilley, Christopher. 2010. *Interpreting Landscapes: Geologies, Topographies, Identities; Explorations in Landscape Phenomenology*. Walnut Creek: Left Coast Press.

VMC – Virtual Museum of Canada. 2014. *Canada Under the Stars*. http://astro-canada.ca/_en/a4102

CHAPTER 3

Aristotelian Streetscapes in the Rise of Modernity

Introduction and Summary

A feminine thread runs through the unfolding of European Neolithic round enclosures into market places of antiquity and the Middle Ages, and into the public sphere of the Enlightenment. As ceremonial sites of public rituals of fertility and solstice renewal, the Neolithic round enclosures were often the communal celebration of the female. Rigorous manual labor was an important component of central-European prehistoric women's behavior, a recent study finds, suggesting an egalitarian communal regime (Macintosh et al. 2017). As cities emerged through the Bronze Age, the conscious notion of a plan most likely emerged as well. Throughout antiquity and the Middle Ages women in Europe acquired a superior stance in the production and sale at town markets of specialized goods such as ales and fabrics. But the deliberate planning of urban precincts, if such took place at all, was the domain of the sovereign and of men.

As opposed to the overt presence of women in urban space since antiquity, it might seem overbearing to seek a feminine expression in the layout of a city's streets. Yet it was Aristotle who stopped short from suggesting a feminine, Dionysian, and masculine, Apollonian, features to urban streetscapes. The two sharply contrasting streetscapes that Aristotle addresses in the *Politics*, are the rigidly elegant and aligned streets reminiscent of the Apollonian disposition of calm and calculated foresight, against the chaotic, crooked streets in poverty-stricken precincts of the city, where

A. Akkerman, *Philosophical Urbanism*,
https://doi.org/10.1007/978-3-030-29085-6_3

Dionysian unpredictability matches the erratic, unplanned narrow lanes and alleyways. The latter type of streets is one that is often ignored or treated with contempt, yet it is Dionysian streetscape where spontaneous human encounter occurs.

As streets and open-air spaces in the ancient and medieval city became places of discourse and exchange of goods, through happy occasions such as frequent street performances and festivals, or through calamities such as fire and the plague, they became also places of communal memory. The mode of communal discourse began to change with the onset of the Little Ice Age (LIA) during the Renaissance and the Enlightenment. Feminine expression in the nascent modern public sphere led to literary and philosophical debates held in salons of women of affluence with their handpicked audiences. The LIA marked the transition from a wide scope of interactions in open-air public space into intellectual discussions in the indoor private space of salons. Emerging from the salons of affluent female salonnières, who led and moderated highbrow discourse through verbal communication with invited guests, was not only the new transformation of the public sphere as a medium for political debate but also the slow demise of the communal place in the city. Be it due to bad weather of the LIA or the atrocious physical condition of streets and open spaces in the city, modernity's gradual ouster of the public sphere into indoor private salons or lecture halls had eroded places of public interchange and communal memory in the city.

During the industrial revolution, the emergent public sphere arising from the voluble salonnières and their audiences was contrasted by the male flâneur engaged in the nonverbal communication of mutual gaze with the urban crowd in shopping arcades and across paved and otherwise improved streetscapes. Ensuing streetscape design based upon surveillance and crowd control further undermined the notion of place in the city as a spontaneously accreted urban niche. As the implicit, unspoken feature of city-form since the new planning of Rome in the sixteenth and seventeenth centuries, urban surveillance came to characterize most of Europe's *new towns*, planned during the Baroque and the Enlightenment on a template of the Ideal City, and in the image of the perceived perfection of the universe.

Even as the cooling of the LIA led to the shielding of segments of public space, the ancient dichotomy pointed out by Aristotle, between the planned, Apollonian streetscape and the slowly accreted and chaotic,

Dionysian streetscape, had persisted intact into early modernity. It was the urban masterplan, Apollonian and masculine, that was usually associated with scale and predictability in affluent precincts, but also with surveillance and control. The chaotic, Dionysian streetscape has been often the place of contempt, of prostitution and crime, continually jostled through modernity by the city's masterplan. It was within such modern urban context that Sigmund Freud reflected on various sections of the city, through its history, and posed the question whether the city could be considered a psychical entity.

In addressing Freud's question, two premises are made here: the one is that the city is, or ought to be, the hub of communal memory; the other premise is that city-form is a gendered hybrid of human bodies and minds, and of volumes and voids. Within such context, the public sphere, as defined by Jürgen Habermas, is a link in a chain of mind-environment progression that has first produced *place*, as a public space of human multitude and communal memory. As the myth of the Citadel came to gradually dominate the myth of the Garden, ever-intensifying spatial surveillance and control in cities had led to the desecration of *place* in later modernity. Where Habermas shows the progress of public space from Classical antiquity into public sphere of the early Enlightenment, Michel Foucault accurately points to the gradual dissolution of public space since the Enlightenment, onto a masculine medium of surveillance in later modernity. With the demise of public space as a place in the city, the Aristotelian dichotomy appears critically vital in the undertaking to restore and revive open-air urban niches of communal memory and public encounter.

3.1 From Urban Prehistory to the Medieval City: Aristotelean Streetscapes as a Paradigm

Mesopotamian cities of Ur, Uruk and Babylon had stone-paved streets dating only shortly after Harappa, and about a millennium before the first two-wheeled chariots hauled by oxen were introduced (Lay 1992: 28). With populations in tens of thousands by the Bronze Age, these Mesopotamian cities, along with Ugarit on the Western Mediterranean coast, had traded with regions further southwest, including Canaan where cities, later noted in the Bible, Shechem, along with Megiddo, Tel Arad or Jericho became crossroads of vital commercial routes for grapes, olives, wheat, or pottery, ferried further east to Egypt.

In Europe and Britain during the later Neolithic trading routes and roads evolved over tracks that had often followed game trails. The first road improvements would have consisted largely of clearing trees and big stones from paths, and as commerce increased, ceremonial pathways leading to the round enclosures had turned into routes connecting communities to markets. Elsewhere, existing tracks were often flattened or widened to accommodate increasing human and animal traffic.

Some of the round enclosures had attained a new feature as "seasonal community gathering places and places of feasting but also as sites of exchange; indeed, as market places [that] were mostly connected to or formed part of places of worship" (Rahmstorf 2018). These designed open spaces of the Neolithic, delineated by ditches or timber poles, could be viewed as the primordial seeds of what more recently Jürgen Habermas coined as the *public sphere*, the realm of deliberate discourse focused on communal issues.

By the late Bronze Age and through the Iron Age, cities in the Mediterranean had processional roads paved with burned bricks and carefully shaped bituminous mortar. Such was the processional walkway leading to the palace at Phaistos, Minoan Crete (c. 2700–1450 BCE), or the half-a-mile long processional way in Babylon, during the seventh century BCE, connecting the city's temple of Marduk to the Gate of Ishtar.

Occasional decoration in Minoan pottery (c. eighteenth–fifteenth century BCE) and frequent motif in Mycenaean pottery (c. seventeenth–twelfth century BCE) had been concentric circles, suggestive of a nightly revolution of the sky, or linear lines radiating from a center, as a solar pattern. Such design evokes spoked wheels on chariots as well as celestial deities (Marinatos 2010: 24–28). Noteworthy, therefore, is the "well-planned radial pattern of built roads covering the immediate hinterland," converging on Mycenae as the center (Castleden 2005: 31; Kelly 1977: 12).

During the Iron Age, at least on the Italian peninsula, two kinds of settlement had evolved almost simultaneously: the one kind of settlement was the hillfort, while the other was the city, *polis* (Bradley 2000: 61). New cities, as urban colonies, were founded on oracular advice, while Iron Age hillforts in Italy as well as in the rest of Europe, were established through warfare, invasion and as military defense posts.

The fundamental distinction between hillforts and *poleis* follows from difference in their respective purposes. It seems fair to say that the *polis* grew and evolved somewhat in a radiating manner round the *agora*, the central civic place, the market and the community hub, descendant of the

Neolithic round enclosure (Hansen 1997). To inland cities, there was seldom an imminent danger of invasion or unwanted intrusion, and the defense of the city needed not be the paramount concern most of the time. The main purpose of the *polis* was to enable commerce, politics and ritual to take place in the open air that was to become gradually the civic space and the conduit of the primeval public sphere.

Hillforts, on the other hand, were built as strongholds, as citadels defined by their delineation through ramparts or palisades, and the temple or the fortress in their midst. In Celtic central Europe, hillforts were the launching grounds of warriors and centers of tribal power, from which surrounding territory was monitored and controlled.

In Classical Greece, other than at Priene and Thurii, Hippodamus is known to have planned upon grid pattern the port of Athens, Piraeus, while the city of Rhodes was laid out by him on a radiating circular plan. There could be little doubt that the Hippodamian urban scheme was a harmonious, ideal city plan that was to correspond to a similarly perceived harmonious communal hierarchy. The Hippodamian ideal-city form expressed by elegantly aligned streets was one category of streetscapes identified by Aristotle in his *Politics.* The other streetscape category presented in Aristotle's urbanist outlook in *Politics* was a street layout celebrating haphazard, twisting lanes resulting from deliberate scheme "to arrange the houses irregularly as husbandmen plant their vines" (Book VII, Ch. 11.). It is compelling to note that the patron of wines and winemaking in ancient Greece was none other than the bisexual god Dionysus, while the patron of city walls and planned colonies was his half-brother, the sun-god Apollo. A discernment Friedrich Nietzsche had made between the Apollonian and the Dionysian dispositions in Greek arts reflects on Aristotle's twofold urbanist outlook, far beyond his own time, as a lasting and relevant call to urban planning.

It is no coincidence that the etymology of the word 'street' points to 'straight' as a source, while the French, 'rue,' is sourced to the Low Latin 'ruga,' meaning 'wrinkle.' Whereas carefully and rigidly aligned streetscapes have always, expressly or subliminally, followed precepts of an ideal city, the tortuous streets of ancient and medieval towns had been the result of centuries-long, gradual urban growth forged by city dwellers, often poverty-stricken homeowners, upon ever-changing conditions of the urban milieu. In Book VII (Ch. 11) of *Politics* Aristotle writes: "A city should therefore adopt both plans [...] The whole town should not be laid out in straight lines, but only certain quarters and regions."

Urbanist notions evolving from ideal urban images of earlier times have been at the heart of Ideal City conceptions through centuries. The very suggestion of the masterplan in modern urban planning too ought to be seen as a version of the ingrained paradigm of the Ideal City. The various notions of an ideal city, and with them associated, perceived ideal societies, possibly originate in Plato's mythical Atlantis (*Timaeus* 24e–25a; *Critias* 108e) and Magnesia (*Laws* VIII, 848d). Rigidity in the layout and the social stratification of Plato's ideal city of Atlantis (Morgan 1998) could be said to have seeped into later ideal urban notions suggesting parity in perfection of physical design and social structure: Platonopolis of Plotinus in the third century CE (O'Meara 2005: 74–75), the City of God of St. Augustine in the fifth century (*The City of God* XV: 26), Utopia of St. Thomas More in the sixteenth century (Dougherty 1986), Christianopolis of Johann Valentin Andreae and the City of the Sun of Tommaso Campanella, both in the seventeenth century (Thompson 1999: 29–33), or the Garden City of Ebenezer Howard at the turn of the twentieth century (Akkerman 2014). The very concept of the masterplan that had emerged in early twentieth century's America with the City Beautiful movement ought to be seen, too, as a Neo-Platonic notion evolving not only from ideal city images of earlier times but indirectly traceable to Plato's Atlantis and Magnesia.

In contrast, Aristotle's urbanist outlook, only a generation after Plato, had addressed not only an ideal city-form of orderly streetscapes but also the distribution of streets laid out irregularly or entirely randomly, with no centralized plan. Aristotle had put forward an urbanist principle adopting aligned as well as erratically distributed streets and open spaces, the ancient philosopher thus having presented an urbanist view not only different from the Platonic ideal city but one incorporating its diametrical opposite. A city-planning standpoint equalizing the significance of an ideal city-form, with that of accreted urban growth of erratic streetscapes, is Aristotle's urbanism, largely ignored through centuries of erudite urban planning. Between Aristotle's view on urbanism and Nietzsche's insight on the Apollonian and Dionysian dispositions in the arts, the Aristotelian challenge to urban planning and design of our own times condenses into the question how, or whether at all, to match in streetscape plan the Apollonian of transparency and expectedness, with the Dionysian of randomness and happenstance. For within our own contemporary city-form reliant on automation, repetitiveness and predictability such approach seems counterintuitive, and almost heretical.

Conscientiously and graciously aligned streetscapes had always followed precepts of architectural beauty, sometimes extending, purposely or subliminally, onto the notion of the ideal city expressive of wisdom, vision and predictability (Rosenau 1983: 42–84). The twisting streets of ancient and medieval towns, on the other hand, had been a writ of quirk and perplexity, the result of gradual streetscape evolution, void of a central plan, coming in spurts of responses over centuries by city dwellers to real-life authentic conditions of urban existence (Frugoni 2005: 6, 45–111). The two different streetscape dispositions under consideration by Aristotle are both presented in his *Politics*, Book VII. Aristotle lauds urban elegance in the city's core of orderly straight streets and aligned open spaces, the embodiment of an ideal city plan, while at the same time endorsing also irregular, twisting streetscapes, as observed in the suburbs near city walls, in precincts considered during antiquity as places of contempt. The lack of any streetscape pattern near the walls was to ensure surprise, albeit only as a defense standard against armed intruders. By adopting both plans, "security and beauty will be combined," writes Aristotle in Book VII, Chap. 11 of his *Politics.*

Aristotle's call to adopt 'both plans' as an urbanist principle strongly resonates today when standardization and uniformity, by necessity, are the prevailing, Apollonian features of city-form. But Aristotle's urbanism is significant to contemporary planning for yet another reason. Whereas cerebral grounds of urbanism and urban planning have been often sought in the philosophical foundations of architecture (Salingaros 1998), objection to such equivocation has been voiced (Goodchild 1990), and to the extent that urban planning can be sourced to notions of the ideal city, its formative origins rest beyond architecture. Nevertheless, as the American architect Robert Venturi had shown, both architecture and urban planning have in common a long history of quest for rational division and utilization of space (Venturi 1966/2008: 88–105).

Aristotle's urbanist reflections present the origins of urban planning and design as primarily related to the open spaces of streetscapes, and to the overall configuration of spaces and edifices within an urban boundary. Quite in contrast to the concerns for elegance and beauty in an object of architecture, Aristotle recognizes the chaotic layout of streetscapes in poverty-stricken areas near city walls as a matter of necessity. These had been the least desirable precincts for residence during antiquity through to early modernity when walls surrounding European cities started to come down. To do away with these poverty-stricken, crime- and vermin-infested neighborhoods vehicular traffic routes were deliberately planned to cut through them. It is all the more remarkable to note Aristotle's advice that the accreted streetscapes not only be preserved but, furthermore, be deliberately fashioned through a ***random*** plan. This extraordinary judgment has been barely noticed by urbanists.

The distinction between the two kinds of streetscape falls well within the dichotomous discernment in the arts, made by Friedrich Nietzsche in the nineteenth century. Within such context, Aristotle's call to adopt 'both plans' as an urbanist principle ought to be seen as an attempt to bring the Dionysian and the Apollonian into a singular urban context. Whether overlooked or avoided, through the long history of city building this advice has remained not much more than a literary note, perhaps precisely due to the often-misguided belief in rationality in urban planning, stemming from the classical architecture model (Kuchenbuch 2015). Countering the rationalist view is the two-millennia-old Aristotelian notion that unification of the two seminal dispositions in the arts of humans within a single urban context ought to be seen as a founding philosophical premise of city-form, and an opening of a conception of urban environment as a dynamic hybrid of minds, bodies and the built form.

3.2 The City as a Psychical Entity: Freud's Urbanist Standpoint on the Historic City

Whereas the Apollonian in city-form represents the myth of the ideal city, the Dionysian, as a counterbalance, epitomizes the paradigm of the Garden in the built environment. Such could be said to be the Aristotelian outlook on city-form. By implication, furthermore, urban space attains the feature of a hybrid medium, not at all independent from mind. Instructive is Benjamin's outlook on buildings in his essay, 'The work of art in the age of mechanical reproduction' (1935/1969). Whereas the Apollonian might be a remote optical view of a building, the Dionysian is the material and immediate contact with building by its occupant:

> Buildings are appropriated in a twofold manner: by use and by perception – or rather, by touch and sight. Such appropriation cannot be understood in terms of the attentive concentration of a tourist before a famous building. On the tactile side there is no counterpart to contemplation on the optical side. Tactile appropriation is accomplished not so much by attention as by habit. As regards architecture, habit determines to a large extent even optical reception. (Benjamin 1935/1969)

The Platonic, or Neo-Platonic outlook on city-form, apparently rejected by Aristotle, stems from an analogy that Plato had drawn between the human soul and the just, ideal city-state. Variants of the city-soul analogy have been sustained as urbanist paradigms over millennia not only by philosophers but also by architects and planners. For most part, the Ideal City, from Plato onward has been the product of the remote, Apollonian outlook, of a thoughtful and cogent plan, rather than an organic concept conceived, growing and maturing from within an actual streetscape.

In psychoanalysis, the Neo-Platonic paradigm seems to have been constructive. In his short treatise *The Ego and the Id* (1923/2010), Sigmund Freud had identified three founding components of the psyche: the *id*, as the set of uncoordinated instinctual dispositions; the *ego*, as the organizing force of the *id*; and the *superego* as a guide and moralizing standard. The negating side of the superego sets a person's conscience, while positive aspirations emanating from the superego yield a person's idealized self-image, the 'ego ideal.'

As three distinct, yet interacting elements of the psychical apparatus, Freud's three psychoanalytic agents suggest correspondence with Plato's tripartite division of the soul (Bennett 1973). The pertinent aspect of Plato's ancient psychology is the fastidious mutuality it puts forward between the human soul and the ideal city. In his search for justice in Plato's *Republic*, Book II (368C–369A), the figurative Socrates employs analogy between the just, ideal city and the soul, presented again in the *Republic*, Book IV (434D–436A). Plato explains that the ideal city-state, the Kallipolis, is divided into three socio-political strata: the producers, farmers and artisans comprising the lowest class; the soldiers defending the city from outside intruders and the police force securing its internal order comprising the auxiliary class; and the guardians who govern the city, comprising the wisdom class.

> The city was believed to be just when the three natural classes within it each did its own, and it was temperate and courageous and wise because of other conditions and states of these classes. [...] We will evaluate an individual as rightly called by the same names as we give in evaluating a city, if he has these same forms in his soul, and the same conditions in them. (*Republic* IV 435B-C)

The similarity between Plato's three components of the soul and Freud's three universal elements of personality is hard to resist (Pappas 1995: 106–111). In Freud, there is also additional consideration of the psyche vs. the city, quite different from Plato's city-soul analogy, and more akin with the notion of urban space as a hybrid molding, as well as molded, by human minds. In *Civilization and Its Discontents* (1929/1994), Freud examines the city as a 'psychical entity':

> Historians tell us that the oldest Rome was the Roma Quadrata, a fenced settlement on the Palatine. Then followed the phase of the Septimontium, a federation of the settlements on the different hills; after that came the city bounded by the Servian wall; and later still, after all the transformations during the periods of the republic and the early Caesars, the city which the Emperor Aurelian surrounded with his walls. [...] Now let us, by a flight of imagination, suppose that Rome is not a human habitation but a psychical entity with a similarly long and copious past—an entity, that is to say, in which nothing that has once come into existence will have passed away and all the earlier phases of development continue to exist alongside the latest one. This would mean that in Rome the palaces of the Caesars and the Septizonium of Septimius Severus would still be rising to their old height on the Palatine and that the castle of S. Angelo would still be carrying on its battlements the beautiful statues which graced it until the siege by the Goths, and so on.[...] (Freud 1929/1994: 6)

In his thought-experiment Freud condenses historic time of a place into a singular event. While the surreal outcome of such an experiment is vividly imaginable, as emphatically shown in Freud's piece, more intriguing is Freud's question whether or under what circumstances could the city be considered a psychical entity.

In asking whether the city can be considered a psychical entity, Freud is getting close to presaging the notion of lifeworld introduced a decade later by his Moravian compatriot, Edmund Husserl. Lifeworld is the range of objects experienced by perceiving subjects, that is, the common cognitive horizon of the perceiving subjects, as the foundation for all shared human experience (Husserl 1936/1970: 142). Freud's thought-experiment is a special case of urban space and historic time, and he concludes that a city, whether or not a psychical entity, is different from mental organism:

> The question may be raised why we chose precisely the past of a city to compare with the past of the mind. The assumption that everything past is preserved holds good even in mental life only on condition that the organ of the mind has remained intact and that its tissues have not been damaged

> by trauma or inflammation. But destructive influences which can be compared to causes of illness like these are never lacking in the history of a city, even if it has had a less chequered past than Rome, and even if, like London, it has hardly ever suffered from the visitations of an enemy. Demolitions and the erection of new buildings occur in the course of the most peaceful development of a city. A city is thus a priori unsuited for a comparison of this sort with a mental organism. (Freud 1929/1994: 6)

The physical history of Rome's built form is notably different from the 'mental organism' of mind and its memory, where built form, by virtue of being a physical space, cannot accommodate two edifices at a single site. The 'mental organism,' on the other hand, most certainly accommodates multitudes of memory records. It would then seem to follow that, indeed, a comparison with a mental organism is unsuitable since in a city "demolitions and replacement of buildings occur." Whereas memory records held in the neurons of an individual's brain, possibly cannot be entirely destroyed during a life, barring a mental disorder, a city's sites, more often than not, possess no perceptible traces of a distant past.

Some 20 years after *Civilization and Its Discontents* had been published the concept of collective memory was advanced by Maurice Halbwachs (1950/1992). Significantly, collective memory has been seen as a building block in the emergence of public sphere (Barash 2016: 39–51), yet neither one of the concepts of collective memory or public sphere has been brought into a concrete context of the built environment.

Taking into consideration the concept of collective memory in Freud's thought-experiment one would recognize a major aspect of truancy. Adopting historical Rome as an example, the community of the city's inhabitants is largely absent in Freud's account. But if we were to extend Freud's thought-experiment to historic Rome *along with* its communities and their collective memories, a different result would obtain. An urban site means one thing to a visitor, but an entirely different thing to someone who has *lived* at the site. Lived space is entirely different from a space gazed upon, even though the space in question is one and the same physical site. Human life, of course, is much shorter than the life of a city, but oral or recorded history of ancestral sites gives them a meaning that to inhabitants is quite different than to passing visitors.

Emergence of the public sphere would inevitably be linked to historic Rome's open spaces. Freud's observation in his urbanist thought-experiment had focused almost exclusively on edifices, designed walls, palaces or other urban constructs that constitute urban volumes, while largely leaving out urban voids. Yet many of the city's historic urban voids were designed as public spaces, those commonly seen as giving rise to the public sphere of early modernity.

Urban voids are addressed by Freud elsewhere. In Freud's other account historic urban voids are not elegant urban public spaces, but tortuous streetscape in the red-light district of the ancient port city of Trieste in northern Italy. Here is how Freud had described his experience of walking through the Dionysian labyrinth of Romanesque streets in the old town section of Trieste:

> As I was walking, one hot summer afternoon, through the deserted streets of a provincial town in Italy which was unknown to me, I found myself in a quarter of whose character I could not long remain in doubt. Nothing but painted women were to be seen at the windows of the small houses, and I hastened to leave the narrow street at the next turning. But after having wandered about for a time without enquiring my way, I suddenly found myself back in the same street, where my presence was now beginning to excite attention. I hurried away once more, only to arrive by another detour at the same place yet a third time. Now, however, a feeling overcame me which I can only describe as uncanny, and I was glad enough to find myself back at the piazza I had left a short while before, without any further voyages of discovery. (Freud 1919/1955: 237)

The place that more candidly than perhaps any other in the city conveys veracity is the place of contempt like the one described by Freud. The confines of the place of urban contempt are – almost by definition—potentially harmful or ugly. Yet it is precisely for the danger, fear and ugliness, that the place of contempt also exudes unpretentious candor. Many places in the city have some repulsive aspects, but—unlike elsewhere in the planned city—the place of contempt displays ugliness as a blunt, defining feature.

The repulsive aspect of a place of contempt is often the apprehension of crime, directly associated with the fear of death, or allied—as Freud intimates—with the fear of castration. The urban void of the place of contempt is a geographic space, bizarre in its ambiance of primordial fierceness of anxiety and dread, very much unlike its urban opposite, the safe, transparent and predictable passageway in a planned streetscape or a carefully designed public space. The place of contempt, Freud's Uncanny, is not one to celebrate, but to recognize as a unique, urban space where the myth of the Ideal City has not taken reign, and where considerable measure of authenticity prevails. Whereas the myth of the Ideal City results in mechanistic ambiance that often appears humanly fraudulent to the fervent observer, the place of contempt and decay in the city appears more as an authentic, Dionysian feature of urban space.

What Freud's two urbanist anecdotes show is that it is a misconception to simply consider urban space as detached from the dynamics of its human context. Furthermore, Freud's two accounts lead to notion of urban space as a hybrid of urban voids and volumes in a continual feedback with human bodies and minds. In this sense, Freud presages Husserl's notion of the lifeworld by alluding to the spatio-temporal scope of juxtaposed urban objects in delineated geographic space of a built environment.

Freud's Trieste account of contemptuous space is echoed in *The Production of Space* by Henri Lefebvre (1974/1992). To Lefebvre urban space is a communal product, which affects spatial perceptions and behavior of people within it. Lefebvre argues that social production of urban space is fundamental to the *re*production of modern society. He sees origins of this feedback progression in the traditional, archaic societies, where the place of contempt, the communal pit, stood also for the communal subconscious. Describing the township of the pre-Roman Greek-inhabited southern part of the Italian Peninsula, Lefebvre observes:

> A pit, originally – a dust hole, a public rubbish dump. Into it were cast trash and filth of every kind, along with those condemned to death, and any newborn baby whose father declined to 'raise' it [...] It connected the city, the space above ground, land-as-soil and land-as-territory, to the hidden, clandestine, subterranean spaces which were those of fertility and death, of the beginning and the end, of birth and burial. (Later, in Christian times, the cemetery would have a comparable function.) The pit was also a passageway through which dead souls could return to the bosom of the earth and then re-emerge and be reborn. As locus of time, of births and tombs, vagina of the nurturing earth-as-mother, dark corridor emerging from the depths, cavern opening to the light, estuary of hidden forces and mouth of the realm of shadows, the mundus terrified as it glorified. In its ambiguity it encompassed the greatest foulness and the greatest purity, life and death, fertility and destruction, horror and fascination. 'Mundus est immundus.' (Lefebvre 1974/1992: 241)

The dual myths of birth and death, of open-air and subterranean existence, of light and darkness, of presence and absence, setting the place of contempt in its communal surroundings, seem to have been formative to the future progression toward urban communities. Thus, Lefebvre asks:

> Might a psychoanalysis of space account for this strange and powerful presence-absence? Undoubtedly, but does it not make more sense, instead of engaging in a posteriori rationalizations of that kind, to envision a slow process of 'historical' secretion, a laying-down and superimposition of strata of interpretation, along with their attendant rites and myths, occurring as the ltaliots localized and focused their fears in the abyssal realm? That a void should be placed at the centre, and indeed at the centre of the conception of the 'world', is surely too strange a fact to be explained solely in terms of psychic realities. (Lefebvre 1974/1992: 242)

The notions of a center and an axis emerge as reckonable qualities in an enigma Lefebvre puts forward, namely, the myth associated with a place of contempt, a void (1974/1992: 171): "In nature, whether organic or inorganic, symmetries (in a plane or about an axis) exist wherever there is bilaterality or duality, left and right, 'reflection', or rotation (in space)." The urban place of contempt too, authentic and real, in a process of Dionysian secretion, counters the Apollonian of the Ideal City in an organic symmetry.

3.3 Historic Streetscapes: Jürgen Habermas and the Public Sphere of Early Modernity

With the onset and progress of the Agricultural Revolution of the Neolithic, accumulation of wealth and storable food in sedentary communities throughout the Fertile Crescent and later in Europe led to the need for defensive structures such as ditches or walls on the perimeters of settlements. The earliest fortifications, large walls primarily, originated in the Fertile Crescent. The oldest wall of which remnants are still extant is one that surrounded the mountainside temple of Göbekli Tepe (tenth–eighth millennium BCE) in the south of present-day Turkey. Protection of granaries seems to be also the reason for walls at many communities of the Harappan Bronze Age civilization in the Indus Valley later in the period 3300–1900 BCE (Pruthi 2004: 137–178).

With abundance of surplus crops, and the ability to conserve and store them, Neolithic settlements in the Fertile crescent grew into Bronze Age cities. The Sumerian city of Uruk (the Biblical Erech, *Genesis* 10: 10) was built c. 4000 BCE, and according to the *Epic of Gilgamesh*, written c. 2100 BCE, King Gilgamesh during his rule c. 2700 had surrounded the city by a series of intricate and beautiful walls (Dalley 2000: 40). Wall inscriptions at Uruk, hailing Gilgamesh, would have been the earliest association between a city and its deified hero.

In Europe, the vast majority of fortified settlements in the Iron Age were small, often with not much more than 50–100 inhabitants, such as those common through Britain. Greece and Italy were more densely populated than the rest of Europe, and settlements there were larger. The fortified Neolithic city of Sesklo in Greece had reached 4000 inhabitants during the sixth millennium BCE. Settlements of the Bronze Age Urnfield culture of central Europe (1300–750 BCE) had sometimes fortified only part of their perimeter, aligning its outer limits with river-bends or swamps as a natural defense. Navigation skill, masculine strength and bravery had transformed into planning, construction and defense of settlements, with the Hero myth in attendance.

Perseus a deified hero defeating monsters and giving rise to the myth of Greece's 12 Olympian gods was also the legendary founder of Mycenae in the northeastern Peloponnese, about 90 kilometers southwest of Athens which was at the time a secondary center of the Mycenaean civilization. Mycenae was a military stronghold which dominated much of southern Greece during the Mycenaean Period of ancient Greece, from about 1600 BCE to about 1100 BCE. At its peak in 1350 BC, Mycenae citadel and the town below had a population of 30,000 and an area of 32 hectares (Thomas and Conant 2009: 1–31).

Huge limestone boulders were used in the construction of the Cyclopean Walls of Mycenae during mid-fourteenth century BCE. In Greek mythology, the Cyclops were a race of giants, each with a single eye in the middle of his forehead. Ancient tradition held that it was the Cyclops who built these walls, as it was thought impossible for men to move the oversized rocks. Only the mythical Cyclopes had the strength to move the enormous boulders (Hurwit 1999: 337).

Associated with hillforts during the late Bronze and early Iron Age, roughly 1100–800 BCE, on the British Isles and through central and western European areas occupied by the Celts, was the Hero myth of the Celts (Harding 2012: 91–118). Navan Fort in County Armagh, Northern Ireland, was said to be founded by the goddess Macha in the seventh century BCE and had been the capital of Conchobar mac Nessa, the mythical king of Ulster. Numerous other mythical heroes of Gaelic Ireland are associated with Navan Fort. Archeological excavation at Navan Fort uncovered large circular enclosure delineated by an embankment and a possible moat. Remains of a mound of earth and stones, possibly raised over graves, are at the center of the enclosure (Harding 2012: 95–112). At least one large round building is believed to have been on the site which in its entirety seems to recall European Neolithic round enclosures (Koch and Minard 2012: 320).

In the south of France Lugdunum, now Lyon, was a Celtic hillfort on the site of which the Roman senator Lucius Munatius Plancus founded in 43 BCE what were to become one of the most important cities in the Roman Empire, the capital of the Roman province of Gallia (Toorians 2008). Lugdunum was named after the Celtic solar god Lugus, as well as a legendary terrestrial hero, Lugh, victorious youth skilled in spear and sling fighting, who became the High King of Ireland (Eastwood 2012: 55).

Hillforts had been the urban source of defensive settlements from which many Iron Age cities had sprung at mounds or high places, often with sudden and sharp elevation rise. Optimal deployment of terrain and topography in Iron Age urban planning could be identified in hillforts (Dyer 2003: 14–46). Fortification usually followed contours of the mound, consisting of one or more lines of ramparts, with palisades or defensive walls, and external ditches. Associated with the need for defense of their communities was also the drive of chieftains to rule adjacent or surrounding territories. Intended for security as well as for projection of control over a surrounding territory, this was also the case of Iron Age hillforts throughout Britain and continental Europe.

Arising through the masculine outlook of warfare or territorial domination had been the myth of the Citadel, identified by Giambattista Vico in his *Scienza Nuova* (Vico 1744/1968: IV- 982). The archetypal Citadel is the Biblical Tower of Babel, a masculine enterprise of supremacy: a manifestation of an attempt to mimic God and to reach to heaven. Erich Auerbach had pointed out Vico's predilection linking the myth of the Citadel with Plato's Ideal City (Auerbach 1949: 110–118), and by implication, with the history of urban planning. In contrast, the primordial parable of the Garden of Eden, with its possibly very earliest version in the *Epic of Gilgamesh*, found a much lesser expression in the planning and building of cities through history.

The feminine Garden found a possible substitute in the multitude of produce and people in the open-air market place, during the Bronze and the Iron Age, and antiquity. The Greek historian, Herodotus of Halicarnassus, noted that in ancient Egypt "the women attend markets and traffic, but the men stay at home and weave" (Herodotus Book II: 35). From the market places at crossroads of trading routes grew open-air markets, and these were likely the other sources to the evolution and growth of towns, a feminine alternative to the archaic forts and fortified settlements. It was the designed urban void of the market place, the *agora*, in the Greek *polis*, dedicated to public gatherings and facilitating communal encounter. A generation after Aristotle, on the north side of the agora in Athens, where the Painted Stoa stood, the founder of Stoicism, Zeno of Citium (333–261 BCE)

> used to discourse pacing up and down the painted colonnade [...] Hither, then, people came henceforth to hear Zeno, and this is why they were known as the men of the Stoa, or Stoics. (Laërtius Vol. II, Book VII, sec. 5)

At the agora of Athens or at nearby streets public argumentation took place in the open air, and the agora was thus turning into a place of civic exchange of ideas, inclusive of questions touching the very core of Athens' political elites (Johnson 2011: 31). Jürgen Habermas views the agora as an important source from which public sphere had originated:

> In the fully developed Greek city-state the sphere of the polis, which was common (*koine*) to the free citizens, was strictly separated from the sphere of the oikos; in the sphere of the oikos, each individual is in his own realm (*idia*). The public life, *bios politikos*, went on in the market place (*agora*), but of course this did not mean that it occurred necessarily only in this specific locale. The public sphere was constituted in discussion (*lexis*), which could also assume the forms of consultation and of sitting in the court of law, as well as in common action (*praxis*), be it the waging of war or competition in athletic games. (Habermas 1962/1991: 3)

Into the Middle Ages, evolution of the market square was conditioned by economic and security concerns, and it was fashioned accordingly. Small settlements often grew close to castles or monasteries, not only for protection but also because of mutual interest in supply of, and demand for goods and services.

From the fifth century, western Europe was plunged into a period of civic disorder that lasted until the ninth century. Faced with the problem of convincing a largely illiterate population of the superiority of a new religion, churches in the Early Middle Ages began staging dramatized versions of Biblical stories, usually on days previously marked for pagan festivities (Young 1926; 1930). The theatrical performances were staged, and often performed by priests at market places or in front of churches. A Christian adaptation of the Roman Saturnalia festival was the secular Feast of Fools, a popular carnival during the Middle Ages, held on January 1 or shortly thereafter. A mock bishop or a pope parodied religious rituals while, for a day, lowly servants exchanged place with high officials. The increasing popularity of open-air secular performances contributed to the weakening of Church authority. By the thirteenth century festivals such as the Feast of Fools had become a lampoon of Christian morality, mocking religious heroes, and subsequently triggering attempts by the Church to prohibit them (Jerman and Weir 2000: 40–47).

From around the twelfth century in Europe, settlements grew into market towns as English and Continental royalty or overlords began granting charters to villages on their land allowing them to conduct occasional or periodic markets. Initially, marketplaces had moveable stalls at squares or at streets in front of homes, with fixed stalls or house front shops becoming later more prevalent. With the demise of feudalism, the protection of suzerains dissipated as well, and merchants had to seek other ways to protect themselves or their property. Guilds offered merchants, for a fee, similar protection and support provided earlier on by kings and overlords. Each occupation had its own guild, which set prices and commercial practices.

Settlements were accorded market town status through custom and practice as royalty and nobility saw market towns attracting people, generating revenue and thus also being able to pay for their protection and land rent. As in antiquity so through the Middle Ages and the Renaissance, market places were the hubs of exchange of goods as well as ideas. Female participation in the markets of towns was mainly as brewers and alewives, but also as artisans and artists. Women were eligible to join guilds, usually by assisting their fathers or husbands in their trade, rather than independently. As a guild member, a widow inherited many of her husband's rights and privileges (James 2017: 236–238).

With increasing populations in urban centers, due to migration and natural increase, physical conditions in cities had deteriorated. Streets covered with the excrement of both people and animals received little attention from the overseeing authority and, saturated with animal entrails and rotting food. Many streets became impassable. As conditions in cites became increasingly unbearable, muckrakers were hired by merchants to clean streets, and some bylaws were instituted to ensure basic public behavior standards such as making it illegal to empty chamber pots out of windows. A recent historical overview of municipal condition in medieval London by *The Telegraph* newspaper stated:

> In 1369, King Edward III demanded that butchers be banned from slaughtering animals within the City of London because of the stench from blood and offal. Rotting meat was commonly dumped in the Thames. Between 1348 and 1665, there were 16 outbreaks of the plague in London. During the Black Death of 1348–49, a third of London's inhabitants died or fled. Other causes of terrible smells in medieval London were the tanneries, where leather would be boiled. People rarely washed or changed their clothes. (*The Telegraph* 2011)

It is for these reasons too that paving, clarity and simplicity in street layout had begun to proliferate in the late Middle Ages, recalling orthogonal or linear layouts that were used in Etruscan, Greek and Roman town planning in antiquity. The architectural and urban aspects of the onset of the LIA were marked by orthogonally and radially planned *new towns* that came on the heels of *bastides*, orthogonally planned fortress towns in southern France and northern Spain during the twelfth century. By the end of the Middle Ages, Spiro Kostof estimated, as many as 1000 *new towns* were founded, not only in Spain and France but also in England, northern Italy and elsewhere in central and western Europe (Kostof 1991: 108–111).

Given the constraints of topography, commercial preferences or private ownership benefits, many of the medieval *new towns*, initially intended to be laid out on a perfectly orthogonal grid plan, ended up deviating considerably from the ideal (Slater 1999: 115–116). It was primarily during the Renaissance that a strikingly novel architectonic style began penetrating the medieval urban ambiance, not only through planned *new towns* but also in the construction of public places in cities. One of the main distinctions of *new towns* and urban *places* of the Renaissance was the increasingly concerted attention to symmetry and geometrical balance, leading to designed optical perspective views utilizing streets as conduits of viewing (Panofsky 1957: 38) (Fig. 3.1).

In the west of Europe, this transition was exemplified by the redesign of Rome during the period 1450–1590. The monumental redesign of Rome was commenced by Pope Nicholas V for the Jubilee celebration of 1450, continued by Alexander VI for the half-millennial celebration of 1500, and completed at end of the century by Pope Sixtus V as part

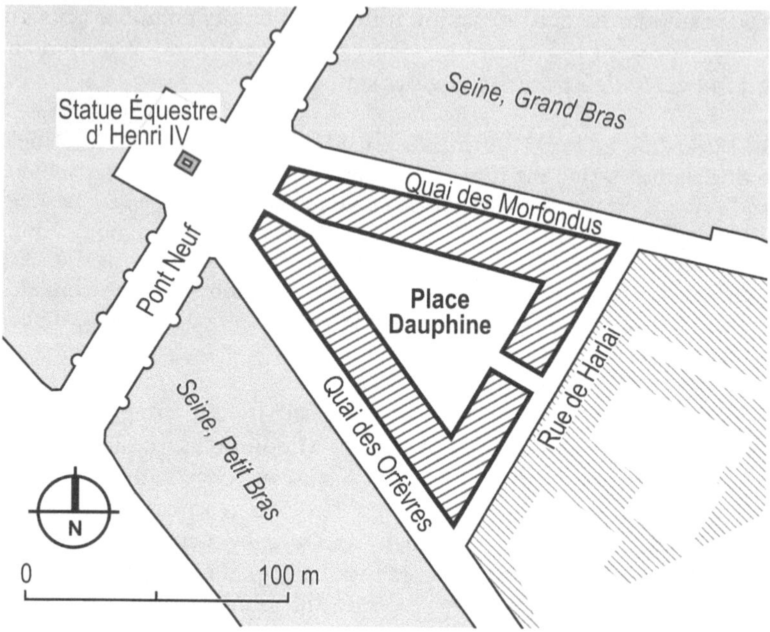

Fig. 3.1 Place Dauphine, Paris, c. 1560. (Drawn by Paul Van Pul)

of his counter-reformation agenda. In order to attract and galvanize large crowds of pilgrims Rome's plan under Nicholas V linked major monuments through a newly built network of straight, radiating roads converging onto a wall fountain, *mostra*, sustained by an ancient aqueduct restored from Rome's time of glory more than a millennium earlier.

The creative spirit behind the new street plan of Nicholas V, and the designer of the *mostra* was the architect and humanist intellectual Leon Battista Alberti (1404–1472), the author of *De re aedificatoria* (1452), a ten-book treatise linking architecture with urban planning. The 12-mile (20 km) aqueduct restored under Nicholas to carry water to the *mostra* came to be known as the *acqua* Vergine, its cusp later expanding into numerous terminal fountains throughout the city of Rome, to sustain residents and pilgrims alike.

To accommodate pilgrims and to alleviate crowding, mainly during Jubilee celebrations, pope Sixtus IV (pp. 1471–1484) built the *Ponte* Sisto and restored other bridges on the Tiber River in Rome. A new Vatican library was built during his pontificate, and a ruined hospital was rebuilt, as were many walls, gates and towers in the city. The inspired force behind these projects was Baccio Pontelli (1450–1492) who also redesigned the Sistine Chapel, pope's official residence, as well as numerous churches throughout the city.

On New Year's Day, 1475, Sixtus IV had issued a decree on the maintenance of Rome, following which major streets of the city were linearly aligned and paved. Later, during the year 1480, prominent buildings were demolished, probably as a security measure to allow clear street views. The official pretext for the demolition was street-paving, and the justification was that such buildings were hindrance to traffic. Roman streetscape design had continued under Pope Alexander VI who had a brand new street, the via Alessandrina, constructed in 1499 to accommodate large numbers of pilgrims in the coming Jubilee year. The aesthetic concern in the newly carved route led to the first deliberate creation of an early modern urban vista by stipulating that buildings have a height of *at least* 16 meters.

During his short papacy, Sixtus V oversaw the final and most significant alterations that had transformed the streetscapes of Rome onto a linear plan, further facilitating traffic flows as well as crowd control. Sixtus V became the early modern force behind the first large-scale urban traffic network, transforming medieval townscapes onto straight Baroque streets, chiefly thanks to judiciousness and skill of Domenico Fontana, an architect and engineer. Supplementing the *mostra* of Alberti and its *acqua* Vergine, were also newly designed monumental fountain wall and its aqueduct restored from Roman antiquity by Giovanni Fontana, Domenico's brother. Under the papacy of Sixtus V the *fontana* dell'Acqua Felice, Domenico's wall fountain, was constructed over the years 1585–88 (Riegl 1908/2010: 160–202). As a terminus of the *acqua* Felice, a 20-mile-long aqueduct, the new fountain enabled the further renewal and growth of Rome, becoming at the same time showpiece of a Church struggling against reformation and early modern science.

The striking street views of the newly redesigned Rome had a lasting impact on Tommaso Campanella, who in 1596 was sent to Rome to stand trial over heresy. In 1599 he was rearrested in southern Italy, tortured and sent to jail again. Campanella's firsthand experience with Domenico Fontana's streetscape design during his brief respite in Rome became another link in a chain of ongoing dynamic feedback between mind and city-form. While in jail, Campanella wrote his famous utopian treatise, *The City of the Sun* (Donno 1981) describing an ideal egalitarian society inhabiting a harmoniously and monumentally designed city.

Campanella was intermittently imprisoned for a very long time and, after enduring suffering and torture by the Inquisition, was set free in 1627. He found refuge in France under the auspices of the Cardinal Armand Richelieu (1585–1642). Campanella had urged the Cardinal to build an ideal city (Yates 1999: 376), and in 1631 the city of Richelieu was, indeed, founded. By 1642 the city's construction as a planned new town was completed on a perfectly rectangular perimeter, with immaculately orthogonal street layout, and large formal gardens (Lancaster 1945). The plan's guiding principle was order and discipline, and on this and similar other plans Michel Foucault made a profound observation:

> [T]here is a central street that divides the rectangle of the town into two rectangles, and then there are other streets, some parallel to and others at right angles to the central street, but at different distances from each other, some closer, others further apart, such that the town is subdivided into rectangles of different sizes, going from the larger to the smaller. [...] In this simple schema I think we find again the disciplinary treatment of multiplicities in space [...] to be constructed and organized according to the triple principle of hierarchy, precise communication of relations of power, and functional effects specific to this distribution [...] Here, it is a case of structuring a space. Discipline belongs to the order of construction. (Foucault 1978/2009: 16–17)

Mutating into no less masculine notion of the masterplan in later modernity, urban planning has transformed from its earlier, overtly mythical calling to rationality onto becoming the subtle rationalist means of ruling authorities to monitor and control the urban crowd. The interim stage of this transformation was driven by the cooling temperatures of the LIA, the decline of the authority of the Church, and the complementary advent of critical thought through the contest between Cartesian rationalism and British empiricism.

3.4 Urbanist Dispositions of René Descartes and John Locke: LIA as a Prelude to Benjamin

Rome's new built form of the early Baroque did not remain confined in its cognitive impact to Campanella's ideal city. The following two centuries saw its mark in the plans for London or Washington, D.C., and some of twentieth-century city-form too took a cue from Rome's ideal-city design

(Berg 2008: 93–116). But its immediate influence was the added thrust in the profusion of planned new towns throughout the Continent, thus inadvertently contributing to the rise of rationalism hastening the decline of Church authority. Furthering Benjamin's stance on urban space as an environ/*mental* hybrid, the monumental, aligned streetscape contrasted with medieval twisted lanes and alleyways. This contrast had forged a link not only to the rationalism of René Descartes but also to the rival empiricism of John Locke. Rationalism, as the search for certainty through absolute reliance on reason, constituted the rejection of Church's monopoly on truth and the affirmation of the autonomy of human mind. The British empiricists, in turn, rejected Descartes' rationalism on the premise that experience through senses is the only valid standard of the search for truth. Both competing approaches in unison largely undermined the supremacy of the Church.

The schism between rationalism and empiricism can be shown as another variety of a feedback between mind and city-form. For both Cartesian rationalism and British empiricism, the latter personified by John Locke, Descartes' fierce empiricist opponent, have their roots in the city-form of their own times. A significant feature in the forming of Descartes' and Locke's epistemologies had been the urban environment, as the ensuing overview of perceptual lineage of their respective outlooks shows. In the case of Descartes, the *sensory experience* of rigorously aligned streetscapes had led to the rationalist call for regimented reasoning. In the case of Locke, the *cerebral advance* that had led to a town plan, a mental image of streetscape plan on a blank slate of paper, had provided analogy to the empiricist claim of cerebral advance from sensory experience upon mind as a blank slate.

Streetscapes have constituted an indubitable perceptual and cerebral stimulus to Descartes' rationalism, and to his co-discovery of coordinate geometry. It is no coincidence that Descartes was not alone in his insight that led to coordinate geometry. Coordinate or analytical geometry was founded independently and simultaneously by Descartes, Girard Desargues (1591–1661) and Pierre de Fermat (1607–1665), at the time of increasing urbanization during the LIA and the profusion of Renaissance planned *new towns*. Geometric investigations during early modernity were related to the perception of optical perspective, initially by the painter and architect Filippo Brunelleschi (1377–1446) at Florence, a city laid out originally on a grid plan.

It is probably also no coincidence that already in the Classical period many geometers and geographers had worked out of Alexandria, a city planned in its entirety largely upon a gridiron pattern. Apollonius of Perga (c. 262–190 BCE), a prolific geometer and author of the book *Conics*, is said to have worked at or near the Library of Alexandria. Eratosthenes (276–195 BCE), the third head of the Library, had provided the foundations for the geographic coordinate grid system, later applied fully by Claudius Ptolemy (100–170 CE), another Alexandrian mathematician. In his work *Geographia* Ptolemy compiled the geographic coordinates of all major sites throughout the world as known during his time, thus formalizing the notion of place, and objectifying geographic space.

The leap made by Descartes was his view that some formal and objective notions, such as those of geometry, are innate ideas. In his *Comments on a Certain Broadsheet* (1647), following a query by the Dutch physician and philosopher, Henricus Regius (1598–1679), Descartes clarified his stance regarding the epistemological status of innate ideas, i.e., archetypal concepts, fundamental to reasoning and common to all people:

> This is the same sense as that in which we say that generosity is 'innate' in certain families, or that certain diseases such as gout or stones are innate in others: it is not so much that the babies of such families suffer from these diseases in their mother's womb, but simply that they are born with certain 'faculty' or tendency to contract them. (Descartes 1647/1988)

To Descartes, certain ideas, such as the concepts of geometry, are innate, and our perception of the environment helps extract them. But unlike the impact of the rigorously aligned streetscapes of Alexandria upon the minds of people like Eratosthenes and Ptolemy, the impact of the linearly planned environment on Descartes seems to have gone beyond geometry. Planned new towns of the Renaissance played a major role not only in his discovery of coordinate geometry but also in the forging of his rationalism. Descartes' account from the onset of the Thirty Years' War, during one of the coldest periods of the LIA, inadvertently posits city-form and weather at a center of his thought. It is his disclosure, in Part Two of the *Discourse on Method* (1637/1968), of a reverie which had occurred to him at the end of a cold autumn day in 1619 near Ulm, Swabia, that reveals Descartes' affinity with clarity of Renaissance planned new towns contrasted with the perplexity of the medieval city-form:

> I was then in Germany, where the occasion of the wars which are not yet over there had called me; and as I was returning to the army from the coronation of the emperor, the onset of winter detained me in quarters where, finding no conversation to divert me and fortunately having no cares or passions to trouble me, I stayed all day shut up alone in a stove-heated room, where I was completely free to converse with myself about my own thoughts. Among the first that occurred to me was the thought that [...] ancient cities which have gradually grown from mere villages into large towns are usually ill-proportioned, compared with those orderly towns which planners lay out as they fancy on level ground. Looking at the buildings of the former individually, you will often find as much art in them, if not more, than in those of the latter; but in view of their arrangement – a tall one here, a small one there – and the way they make streets crooked and irregular, you would say it is chance, rather than the will of men using reason, that placed them so. (Descartes 1637/1988: 116)

Holland, Descartes' repeated destination between 1618 and 1648 was a major arena for Europe's planned *new towns*. But in his frequent travels, Descartes would encounter planned *new towns* also in France and Germany. Two early Renaissance planned *new towns* lie on a direct path between Paris and Ulm, the route of many protestant refugees before and during the Thirty Years' War. The one planned town is Vitry-le-François, founded in 1544, and the other is Freudenstadt, founded in 1599. Vitry-le-François was built in 1544–1545 at the behest of Francis I, king of France (r. 1515–1547) as a modern city, on site of a town destroyed in war hostilities. A strict gridiron plan of four large street-blocks, each including four smaller blocks, guided the design of the Italian military engineer and architect Girolamo Marini (1490–1553) (Fig. 3.2). Freudenstadt, Swabia, too was built on an orthogonal pattern (Fig. 3.3). Designed by Heinrich Schickhardt as a bastion for Friedrich I, the Duke of Württemberg, Freudenstadt was to house many of the 11,000 protestant refugees from Catholic Austria, who fled in the period 1598–1605, and were employed in the duke's nearby Christophtal copper and silver mines (Lewis 2016: 58). At the beginning of the Thirty Years' War, the time of Descartes' account, both Vitry and Freudenstadt were regional centers of Protestant religion and refuge (Bouchot 1878; Braunfels 1988: 149–150).

It was in all likelihood Freudenstadt which Descartes had in mind on November 10, 1619, the day in the stove-heated room, when he recalled "those orderly towns." The grid plan of the town had traced the Bavarian board game Mühle which, in the historical context of frequent peasant

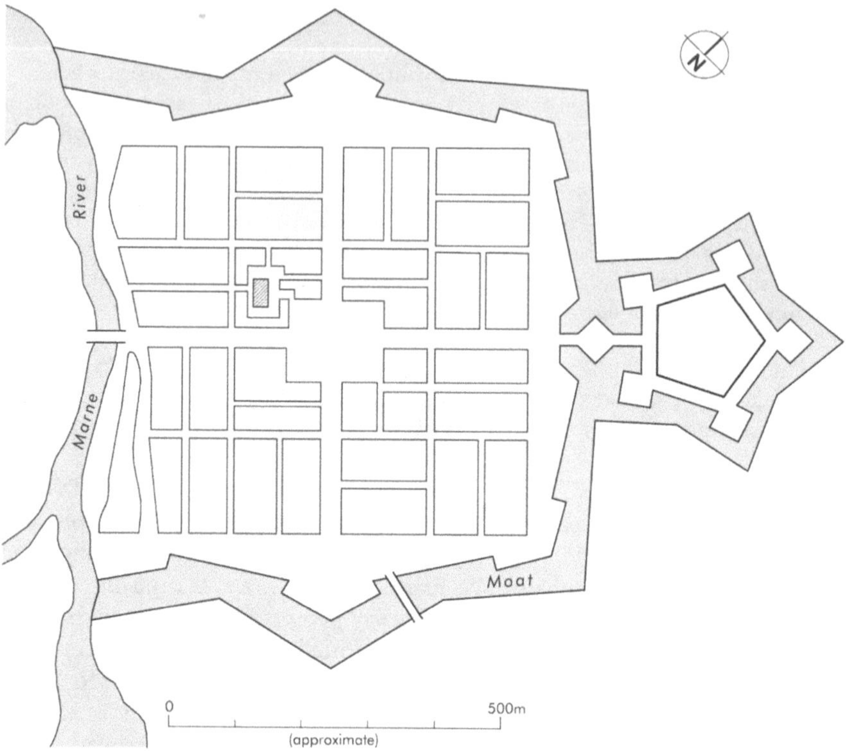

Fig. 3.2 Plan of Vitry-le-François, c. 1545. (Drawn by Paul Van Pul)

uprising at the time, was not frivolous, but more likely intended to promote surveillance of the miners and their families. This space was intended to be the container for people expected to behave in a predictable manner, and under control.

The objective of surveillance and crowd control in Renaissance urban planning was consistent with utopian schemes, whereby the carefully laid out streets of new towns and public places were the fulfillment of an urban dream: an ideal city that reflects the perfection of cosmos and wisdom of the Sovereign through the symmetry and harmonious geometry of edifices, streets and squares (Dennis 1988: 43–51; Donno 1981: 14). Other meticulously designed Renaissance *new towns* of Germany and France, with which Descartes would have been familiar through his travels, would likely provide the same or similar perceptual impetus.

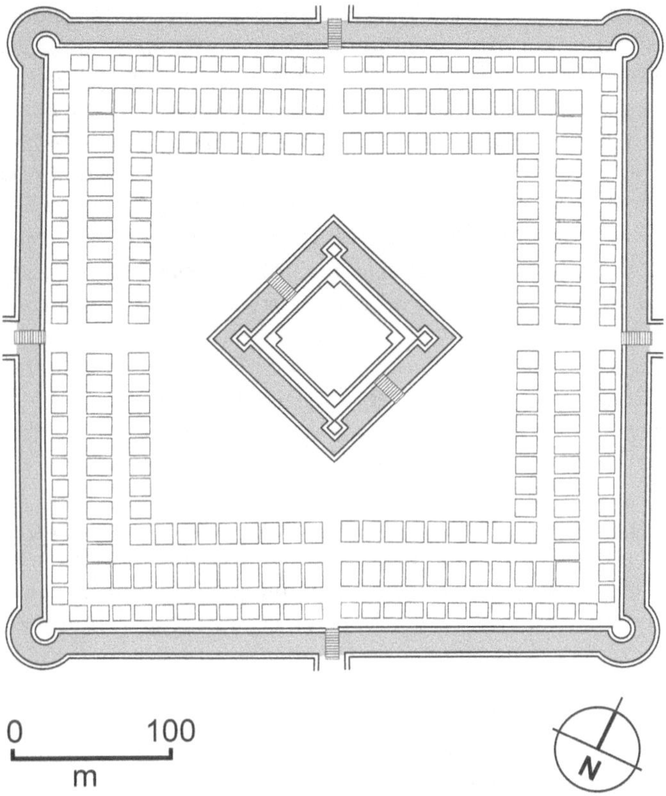

Fig. 3.3 Plan of Freudenstadt, 1599. (Drawn by Paul Van Pul)

It is from within this urban ambiance that modern rationalism was ushered by Descartes. Scientific rigor and technology that had emerged from Cartesian thought had led later, in turn, to the modern myth of the Rational City, a vision of fully mechanized, automated and well-functioning urban metropolis. Urbanist concepts such as the City in the Park of Le Corbusier, or even the grid pattern expediting automobile access in many American cities have been seen by some as the urban epitomes of Cartesian rationalism (Frampton 1992: 180; Gandelsonas 1998: 45–58).

Historical coincidence had it that less than half a century after Descartes' *Discourse*, John Locke became the chief planner for the Province of Carolina, an English colony in America. Modern rationalism propounded by René Descartes had come to face rejection by the British empiricism at whose birth stood none other than John Locke. Yet the conflicting doctrines of Descartes and Locke, two founding fathers of modern thought, have been nurtured by philosophical ideas that had emerged from their own environmental contexts, within which city-form had played a major role.

It was in his capacity as Carolina's Chief Planner that Locke had drafted in 1670 the 'Grand Modell for the Province of Carolina' on behalf of King Charles II of England, and at the behest of Anthony Ashley Cooper, the First Earl of Shaftesbury (1621–1683), one of the Lords Proprietors of Carolina. Locke's 'Grand Modell' included a constitution and detailed settlement plan for the Province of Carolina, with a gridiron plan of Charles Town, the capital city. The capital was indeed founded in 1670, built from scratch on an Atlantic coastline plane at the confluence of two rivers, Ashley and Cooper, according to Locke's plan and following Shaftesbury's specifications (Fig. 3.4). In 1719 it was renamed Charleston, becoming the capital of the new State of South Carolina.

Several years after the founding of Charles Town Locke had commenced writing *An Essay Concerning Human Understanding*, first published in 1689. It was in his *Essay* where Locke introduced his empiricist notion of clean slate, *tabula rasa*, as the initial condition of human mind. Only perceptual experience leaves its traces in the mind, argued Locke, and only based on experience reasoning emerges, writes the father of British Empiricism in his *Essay* (Book II, Ch. 1, sec. 2):

> Let us then suppose the Mind to be, as we say, white Paper, void of all Characters, without any *Ideas*; How comes it to be furnished? Whence comes it by that vast store, which the busy and boundless Fancy of man has painted on it, with an almost endless variety? Whence has it all the materials of Reason and Knowledge? To this I answer, in one word, From *Experience*: In that, all our Knowledge is founded; and from that it ultimately derives it self. (Locke 1689/2005: 66)

The plan-making background of Locke seems to have played a considerable role in his empiricism. The analogy of a town-plan drawn on a white paper according to which an actual town of bricks and mortar is built on a virgin ground, on the one hand, with mind as a blank slate onto which observational experience carves impressions, on the other hand, is almost inescapable.

The two early founders of the Enlightenment, philosophical adversaries who fathered European rationalism and British empiricism, source their

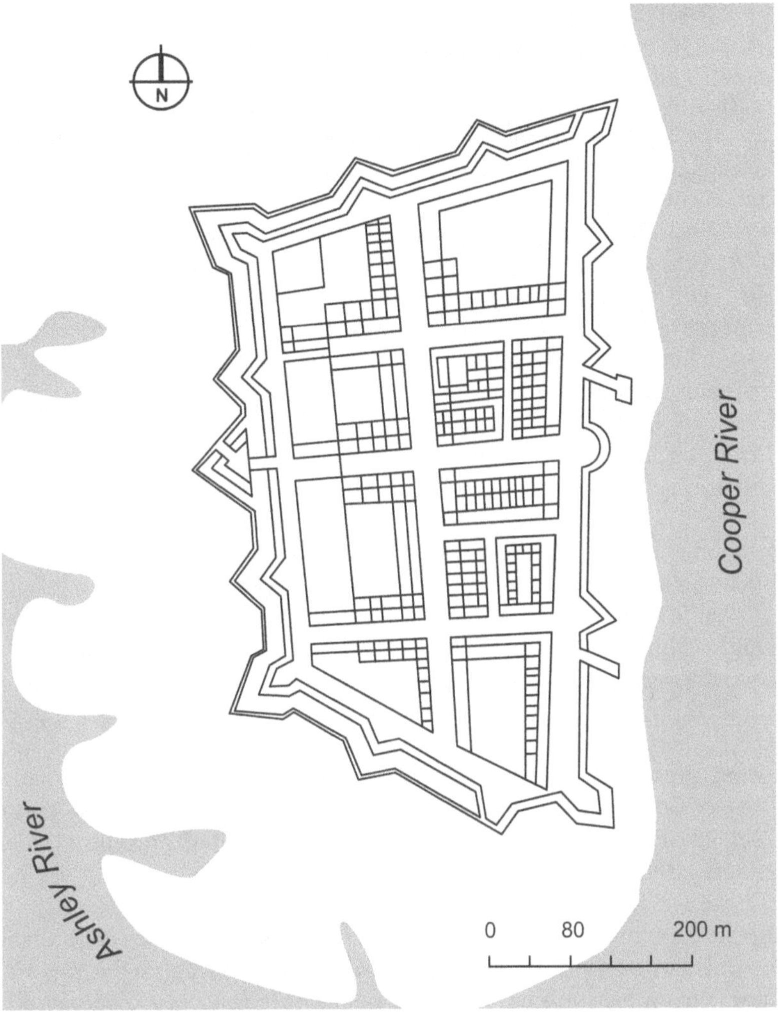

Fig. 3.4 Plan of Charles Town, after John Locke, c. 1670. (Drawn by Paul Van Pul)

reciprocally contrarian arguments upon a common origin. Their mutually discordant reasoning arises from a strikingly similar conceptual and perceptual urbanist base: the planned new town of Descartes' time, and Locke's own 'Modell' for Charles Town (Akkerman 2020: 37–42).

3.5 Public Sphere and Public Surveillance: Habermas Versus Foucault

Descartes was led through the sensory experience of a regimented urban streetscape to the rationalist stance of regimented reasoning. In a kind of reverse analogy, Locke came to insist on cerebral progression corresponding to the advance from a town plan on a blank slate of paper onto actual construction of a town. The planned built form of a town is in the midst of a mutual complementarity between Locke's empiricism and Descartes' rationalism. The town plan drawn on a 'white Paper' had been the *cerebral* imprint of the empiricist claim of progression from sensory experience upon mind as a blank slate; the planned town, as the *perceptual* expression of "the will of men using reason" has been the tenor of Descartes' rationalist posture of rigor in thought, based on clarity and distinctness of concepts.

Locke's 'Grand Modell' came on the heels of intensive urban planning deliberations in London following the city's devastation in the Great Fire of 1666. This effort engaged the best minds of the Royal Society, Sir Christopher Wren and Robert Boyle among them, both of whom submitted proposals for the Plan for London. As Locke was elected a fellow of the Royal Society in 1668, he would be privy to the urbanist discourse at the highest level. The larger context of Locke's undertaking was set by the return of King Charles II to England in 1660, following a 9-year exile and turmoil of the Third English Civil War (1649–1651) between supporters of the Parliament, and the Royalists. This was strife in the public sphere between those who believed in the representation of commoners in the sharing of political power, particularly the enforcement and levy of taxes on behalf of the state, against those who advocated absolutist rule by monarchy, deferring to the divine right of kings.

Locke's 'Grand Modell' of the 1670s has been thus seen also as a stage in Locke's political development, from his belief in benevolent absolutism to his later support of liberal ideas. Locke had traversed a vast moral distance, from his support of slavery in the *Fundamental Constitutions of Carolina* in 1669 to endorsement of the liberalism of constitutional monarchy in England following the removal from throne of the Catholic King James II in 1688 (Wilson 2016: 25–57). Much of this 20-year long transformation could almost certainly be attributed to the emerging public sphere of the early Enlightenment. In the founding of British empiricism, the 'Grand Modell,' however, looms much larger in the imagery of the source to Locke's notion of mind at birth as an empty receptor.

The all-encompassing philosophical doctrine of Locke's empiricism, as emanating from an urbanist theme, is a supposition reinforced also by the early source to empiricist thought in Francis Bacon's *New Atlantis* (1627/1909) where scientific learning and discovery are pursued within the confines of an ideal city, Bensalem. Envisaged on a mythical island, Bensalem, much like Plato's Atlantis or Thomas More's Amaurot on the island of Utopia (More 1516/2016), all imply unambiguous new beginnings of an ideal society in an ideal city, in isolation from the rest of the world. Behind Locke's 'Grand Modell' that was to be pursued in the New World, there too is the desire to build an ideal community. Bacon's Bensalem of the *New Atlantis* appears to be the inspiration, a new beginning, a blank slate from which an ideal community springs, as well as allegorical source of Locke's empiricism (Anstey 2011: 49; Wilson 2016: 1–30, 99–102).

Somewhat reminiscent of the utopian Bensalem in Bacon's *New Atlantis*, Habermas considers the public sphere to have evolved during the Enlightenment into a medium of rational discourse (cf. Murray 1987: 10–11). Habermas sees the rise of modern rationalism, in particular, as a critical marker in the continuum of evolution of the public sphere from antiquity to modernity, giving credit to civic virtue in the founding of the modern middle class, the bourgeois urban community:

> [T]he polis provided an open field for honorable distinction: citizens interacted as equals with equals, but each did his best to excel. The virtues, whose catalogue was codified by Aristotle, were ones whose test lies in the public sphere and there alone receive recognition. [...T]hroughout the Middle Ages the categories of the public and the private and of the public sphere understood as *res publica* were passed on [...] They found a renewed application with the rise of the modern state and of that sphere of civil society separated from it. (Habermas 1962/1991: 3–4)

Rational thought has been associated also with urban planning throughout much of history, from early antiquity to modernity. Yet urban planning and Habermas' public sphere, through considerable segments of history were not at all congruent but quite apart from one another. The truth is that from antiquity through to early modernity urban planning had been largely driven by the myth of Ideal City in the geometric image of the perfection of the universe, whereas early public sphere as a communal place of encounter was driven by religion, cult, theater performance and the market square.

The planned town has been, since the Renaissance, largely the harbinger of comfort and prosperity to the middle class. Inconspicuously, it became also the messenger of subtle control, subjecting the working poor and those of low social status, in particular, to curtailment and surveillance. It was thus the abandoned quarter of the city, left untouched by authorities, that for centuries, attained an air of danger as well as freedom. Three hundred years after Descartes, Walter Benjamin mused in Convolute P of the *Arcades Project*:

> And then the timeless little squares that suddenly are there, and to which no name attaches. They have not been the object of careful planning, like the Place Vendome or the Place des Greves, and do not enjoy the patronage of world history, but owe their existence to houses that have slowly, sleepily, belatedly assembled in response to the summons of the century [...] For what do we know of street corners, curbstones, the architecture of the pavements – we who have never felt heat, filth, and the edges of stones beneath our naked soles, and have never scrutinized the uneven placement of the paving stones with an eye toward bedding on them. (Benjamin 1940/1999: 516–517)

Administration of cities, from antiquity to modernity, had focused on the poverty-stricken parts of the urban community only when they threatened the authority, or in response to threats by the middle-class mainstream. Over this lengthy period, cities have evolved in response to the voices heard in the affairs of the city by those of the imperial authority, of the burghers and later the bourgeoisie and the rest of the middle class. The poor and the fringes of the urban community had a voice in the affairs of their city, only to the extent that their own actions, miniscule and often illegal, carried impact upon the city's streetscapes.

It is rather striking that in spite of the disparity between rationalism and empiricism, the planned town was the object of praise and a cerebral template to the philosophical doctrines of both Descartes and Locke. It is also true that since at least Plato's Atlantis the planned towns have been variations of the myth of the Ideal City. Furthermore, as critical thought was taking hold in Europe north of the Alps, leading to the rise of the scientific revolution, the Baroque unfolding of city-form, exemplified by the new urban plan of Rome or the many other planned *new towns*, such as Richelieu, had proceeded along the slow decline of Church authority. As of the redesign of Rome in the fifteenth and sixteenth centuries, through the late Renaissance towns of Descartes' time, spectacle and crowd control have hybridized in the early modern streetscape.

While Rome's noble streetscapes were intended to counter Reformation and discontent with the Church among commoners and pilgrims from the countryside, in the north of the Italian peninsula an outright secular trend was taking place among aristocracy. Women of distinction and wealth were opening salons to invited guests for periodic events where poetry and literature was read, and where critical opinions were argued by participants. Throughout the seventeenth century the habit of aristocratic women to host invited literary gatherings at which towering figures of literature and the arts were present, had spread into France and then gained momentum in the rest of Europe. This confined literary public sphere had "transformed into the political public," ultimately contributing to the French Revolution of 1789 (Goodman 1994: 280).

In Paris, from the eve of the French Revolution and into the nineteenth century, the high-class female salonnière continued as the hostess of sophisticated verbal articulation of political discourse within the literary public sphere. Her urban complement and male opposite was the flâneur, a significant human element in the milieu of the commoners' urban space. The feminist and political aspects of the salons suggest a flair of defiance against the paternalistic mainstream where the sophisticated side of the salon debates gave rise to action-inspiring conversations among members of the high society. The action-laden dialogues had spread into the broader public sphere of newspapers, journals, masonic lodges, coffee houses and reading clubs, whereby the urban middle class was able to engage in debates involving contrasting arguments on issues of the day. It was the wide appeal to take part in the nascent public sphere that was instrumental in the French Revolution and, subsequently, also in the advent of open society and democratic rule (Habermas 1962/1991: 31–42).

In his later work Jürgen Habermas had introduced the notion, *communicative action*, in an attempt to generalize the pattern of discourse in the public sphere, leading to coordinated attainment of a common goal:

> [T]he concept of communicative action refers to the interaction of at least two subjects capable of speech and action who establish interpersonal relations (whether by verbal or by extraverbal means). The actors seek to reach an understanding about the action situation and their plans of action in order to coordinate their actions by way of agreement. (Habermas 1981/1984: 86)

Contrasting the public sphere of communicative action was the quiet surveillance associated with the persona of the flâneur, a unique form of nonverbal communication taking place in urban environment saturated by the crowd. The gaze, as a silent interaction within the urban crowd, or between the flâneur and the urban crowd, had no subversive connotation at first, but it too constituted novel configuration within the industrializing urban environment. The one aspect of novelty were the arcades, as shielded pedestrian passageways, along with occasional open-air streetscapes that were gradually becoming subject to municipal improvements. The other significant aspect of the flâneur and the crowd within this newly carved urban space was the single-pronged notion of the stranger.

Urban planning had a role to play in the emergence of nonverbal communication, surveillance in particular, within the public space of nineteenth century's European metropolis. To Habermas, to whom lifeworld involves mainly, or only, language and communication (Habermas 1988/1999), urban planning, along with the rest of science, technology, bureaucracy and market economy, would presumably play the role of instrumental rationality hampering communicative action, whose vitality is necessary to secure openness in a *mass* society. But that would be an understatement: urban planning of the eighteenth and nineteenth centuries came to serve the controlling authority, and surveillance was the primary, preventative, means of crowd control.

Surveillance and crowd control were the rather explicit functions of the *Baupolizei*, the Royal-Prussian urban planning police that hired (in 1858), then fired (in 1861), James Hobrecht. The father of the Hobrecht-Plan of Berlin wished to retain some historical heritage streetscapes within the newly expanding city. But the tortuous, narrow streets of medieval Berlin would run against the way of thinking of Hobrecht's superiors. No cause for Hobrecht's dismissal was given, but the discerning planner's urban conservation views run counter to the concern for security that begged street design for expectedness and transparency to eliminate attempts at insurgency in the restive nineteenth-century Europe.

Expectedness and visual oversight is exemplary in the urban concept of Claude-Nicolas Ledoux (1736–1806) in his volume *L'Architecture considérée sous le rapport de l'art, des moeurs et de la législation*, published in 1804 in Paris. Ledoux's plan for a workers' settlement, Chaux, was based on the project of a small mining colony, the Royal Saltworks at Arc-et-Senans, which he designed during 1773–79 for Louis XIV on a tract near the Swiss border. Ledoux's proposal is dominated by the *Administrator's House*, a spectacular edifice with a glaring entrance portico upon which images of gushing, saline water, are carved. The entire Chaux urban design scheme is otherwise strictly functional, set on a sym-

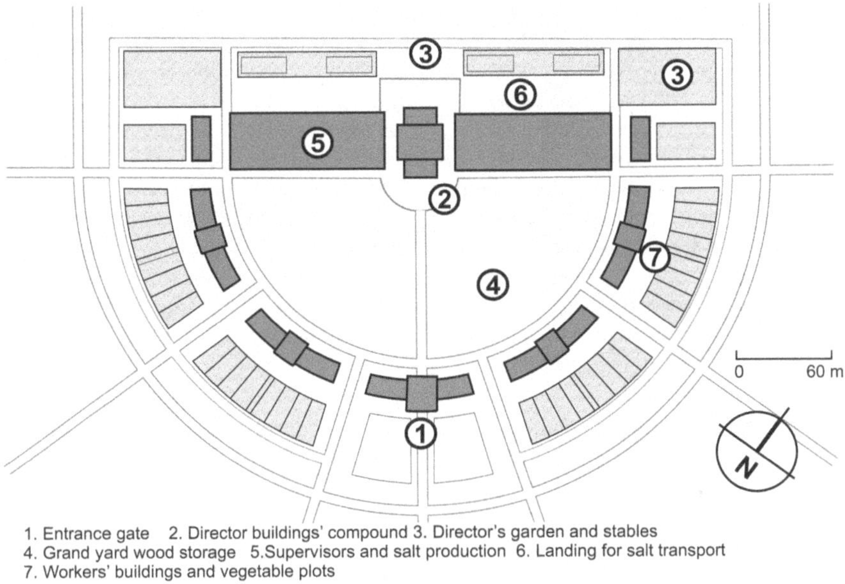

Fig. 3.5 Plan of Chaux, after Claude-Nicolas Ledoux, c. 1774. (Drawn by Paul Van Pul)

metrical, rigidly circular plan with the administrator's house at the exact center, adhering to the perfection of Cartesian rigor (Hersey 2000: 55–58). Ledoux here attempted to create an explicit cosmic paradigm, furthering his own archetypal city-universe analogy, inaccurate as it were, of a planetary system, faithful to Enlightenment's Age of Reason.

A semi-circular layout on one side of the director's house was intended to contain guardrooms, a prison and a forge, as well as quarters for carpenters, coopers and other laborers (Fig. 3.5). The semi-circle on the other side of the director's house was to consist of the saltworks themselves, where drying ovens, heating pots, supervisors' quarters, and the salt stores were to be located. In his book, *Discipline and Punish* (*Surveiller et punir*, 1975/1995), Michel Foucault explains the reasoning as well as the impact of Ledoux's project:

> The perfect disciplinary apparatus would make it possible for a single gaze to see everything constantly. A central point would be both the source of light illuminating everything and a locus of convergence for everything that

> must be known: a perfect eye that nothing would escape and a center toward which all gazes would be turned. This is what Ledoux had imagined when he built Arc-et-Senans; all the buildings were to be arranged in a circle, opening on the inside, at the center of which a high construction was to house the administrative functions of management, the policing functions of surveillance, the economic functions of control and checking, the religious functions of encouraging obedience and work; from here all orders would come, all activities would be recorded, all offences perceived and judged; and this would be done immediately with no other aid than exact geometry. Among all the reasons for the prestige that was accorded, in the second half of the eighteenth century, to circular architecture, one must no doubt include the fact that it expressed a certain political utopia. (Foucault 1975/1995: 169)

A century after the plan of Chaux, a set of diagrammatic plans of the Garden City by Ebenezer Howard (1898), considered a founder of modern urban planning, showed considerable affinity with Ledoux's plan. A dozen other geometric radial and concentric plans of utopian cities were produced in the period between Ledoux and Howard, most or all of them inspired, directly or indirectly, by the Panopticon prison project of the British philosopher and social reformer Jeremy Bentham (1747–1832).

The source of Bentham's penitentiary project (Bentham 1791/1995) was his younger brother Samuel (1757–1831), who in 1786, while employed in Russia, prepared a plan for a circular labor compound with a central surveillance hub, as a means of allowing him to oversee the activities of a large, unskilled workforce of peasants. Samuel had likely borrowed from Ledoux the principle of a single director overseeing the salt-mine workers through one-way vision, effectively securing constant subordination. In 1797 Bentham published *Pauper Management Improved: Particularly by Means of an Application of the Panopticon Principle of Construction*. Bentham's saw the architectural concept of surveillance as applicable not only to prisons but also to hospital units or mental asylums where patients require constant monitoring. A direct line of reasoning, from Ledoux to Bentham had been in fact suggested as well (Bender 1987: 18–22).

In Howard's exceedingly detailed scheme of the Garden City, there is no mention of a single police station. From Howard's Panopticon-like diagrams it is more likely that the police is watching, and watching *everybody*, from a concealed central location. Along with the subtitle to his Garden City handbook, *A Peaceful Path to a Real Reform*, Howard's layout chillingly resonates of Bentham's own outline of the principle of the Panopticon plan:

> [I]t should be clearly understood what circumstances are, and what are not, essential to the plan. The essence of it consists, then, in the centrality of the inspector's situation, combined with the well-known and most effectual contrivances for seeing without being seen [...T]he person to be inspected should always feel themselves as if under inspection, at least as standing a great chance of being so [...] (Bentham 1791/1995)

To Jane Jacobs, the acclaimed urban planning reformer of the late twentieth century, "Howard set spinning powerful and city-destroying ideas" in his Garden City scheme (Jacobs 1961: 18). In her bestselling book, *The Death and Life of Great American Cities*, Jacobs derides Howard's aligned streetscapes as rigidly planned and alienating. While Jacobs' criticism is quite appropriate, more accurate suggestion is that Howard's Garden City diagrams are evidently guided by perceived need for surveillance, just like the plans of Freudenstadt or Chaux, for example, but also similar to Jacobs' own prescriptive advice:

> [T]he sidewalk must have users on it fairly continuously, both to add to the number of effective eyes on the street and to induce the people in buildings along the street to watch the sidewalks in sufficient numbers. [sic] (Jacobs 1961: 35)

Between Howard and Jacobs, the two urbanist adversaries, implicit accord exists in support for what Foucault had called "the disciplinary society of surveillance." In his incisive observation Foucault points to the ultimate panopticism where everyone watches everyone in an Apollonian mechanism of nonverbal means of control:

> [...] it is the apparatus as a whole that produces "power" and distributes individuals in this permanent and continuous field. This enables the disciplinary power to be both absolutely indiscreet, since it is everywhere and always alert, since by its very principle it leaves no zone of shade and constantly supervises the very individuals who are entrusted with the task of supervising; and absolutely "discreet," for it functions permanently and largely in silence. Discipline makes possible the operation of a relational power that sustains itself by its own mechanism and which, for the spectacle of public events, substitutes the uninterrupted play of calculated gazes. (Foucault 1975/1995: 170)

Adhering to the tenets of panopticism, Foucault points out, had been the design of urban space exercised in the late Enlightenment by regimentation in planning across France, from Arc-et-Senans in the east, by Ledoux, to Nantes in the west by Vigné de Vigny. The appeal to authority in the surveillance and control of urban space through regimented urban design was the belief that it, like the Panopticon, "creates a consciousness of permanent visibility as a form of power, where no bars, chains, and heavy locks are necessary for domination any more" (Allmer 2011). The facility of spatial surveillance was thus accorded increasing significance by political elites as well as local administrative authorities:

> [T]actics of power fulfils three criteria: firstly, to obtain the exercise of power at the lowest possible cost (economically, by the low expenditure it involves; politically, by its discretion, its low exteriorization, its relative invisibility, the little resistance it arouses); secondly, to bring the effects of this social power to their maximum intensity and to extend them as far as possible, without either failure or interval; thirdly, to link this 'economic' growth of power with the output of the apparatuses (educational, military, industrial or medical) within which it is exercised; in short, to increase both the docility and the utility of all elements of the system. (Foucault 1975/1995: 218)

The attempt to monitor and control urban populations for reasons of safety and security has been among the causes that had led to the demise of public space. What came to be called *the loss of place* in the latter half of the twentieth century has consisted of efforts to transform urban squares into intersections controlled by traffic lights, to convert urban pathways and passageways into clear fields of vision to drivers, and to turn streets into 'naturally' monitored, 'defensible spaces' to pedestrians (Newman 1996: 9–30). Where Habermas concisely addressed the ascent of public space from Greek antiquity into public sphere of the early Enlightenment, Foucault accurately showed the gradual purge of public space since the Enlightenment, onto a medium of surveillance in later modernity. Contrasting Foucault's observation of increasing urban surveillance and control is the ancient Aristotelian dichotomy of streetscapes that re-emerges as a vital challenge, as well as a promise, in the endeavor to restore and revive open-air urban niches of communal memory and public encounter.

Bibliography

Akkerman, Abraham. 2014. Platonic Myth and Urban Space: City-Form as an Allegory. *University of Toronto Quarterly* 83 (4): 757–779.

———. 2020. *The Urban Archetypes of Jane Jacobs and Ebenezer Howard: Contradiction and Meaning in City Form*. Toronto/Buffalo: University of Toronto Press.

Allmer, Thomas. 2011. Critical Surveillance Studies in the Information Society. *TripleC* 9 (2): 566–592.

Anstey, Peter R. 2011. *John Locke and Natural Philosophy*. Oxford/New York: Oxford University Press.

Auerbach, Erich. 1949. Vico and Aesthetic Historism. *Journal of Aesthetics and Art Criticism* 8 (2): 110–118.

Bacon, Sir Francis. 1627/1909. *The New Atlantis*. New York: P.F. Collier & Son.

Barash, Jeffrey Andrew. 2016. *Collective Memory and the Historical Past*. London/Chicago: University of Chicago Press.

Bender, John. 1987. *Imagining the Penitentiary: Fiction and the Architecture of Mind in Eighteenth Century England*. Chicago/London: University of Chicago Press.

Benjamin, Walter. 1935/2019. The Work of Art in the Age of Mechanical Reproduction. In *Illuminations: Essays and Reflections*, ed. Hannah Arendt, 166–195. Boston/New York: Mariner Books.

———. 1940/1999. *The Arcades Project*. Trans. Howard Eiland and Kevin McLaughlin, Rolf Tiedemann (ed.). Cambridge, MA/London: Belknap Press of Harvard University.

Bennett, Simon. 1973. Plato and Freud: The Mind in Conflict and the Mind in Dialogue. *The Psychoanalytic Quarterly* 42 (1): 91–122.

Bentham, Jeremy. 1791/1995. Panopticon or the Inspection House. In *The Panopticon Writings*, ed. Miran Bozovic, 29–95. London: Verso.

Berg, Scott W. 2008. *Grand Avenues: The Story of Pierre Charles L'Enfant, the French Visionary Who Designed Washington, D.C.* New York: Vintage Books.

Bouchot, Henri. 1878. *La société à Vitry-Le-François aux XVIIe et XVIIIe siècles, d'après des documents inédits*. Vitry-Le-François: Pessez.

Bradley, Guy. 2000. *Ancient Umbria: State, Culture, and Identity in Central Italy from the Iron Age to the Augustan Era*. Oxford/New York: Oxford University Press.

Braunfels, Wolfgang. 1988. *Urban Design in Western Europe: Regime and Architecture, 900–1900*. Trans. Kenneth J. Northcott from the German Original. Chicago/London: University of Chicago Press.

Castleden, Rodney. 2005. *The Mycenaeans (Peoples of the Ancient World)*. Abingdon: Taylor & Francis.

Dalley, Stephanie. 2000. *Myths from Mesopotamia: Creation, the Flood, Gilgamesh, and Others*. Oxford/New York: Oxford University Press.
Dennis, Michael. 1988. *Court and Garden: From the French Hôtel to the City of Modern Architecture*. Cambridge, MA: MIT Press.
Descartes, René. 1637/1988. *Discourse on Method*. In *Descartes' Selected Philosophical Writings*, ed. and trans. J. Cottingham, Robert Stoothoff, and Dugald Murdoch. Cambridge: Cambridge University Press.
———. 1647/1988. Comments on a Certain Broadsheet. In *Descartes: Selected Philosophical Writings*, ed. Cottingham, John, Robert Stoothoff, and Dugald Murdoch, 213–218. Cambridge/New York: Cambridge University Press.
Donno, Daniel J. 1981. The City of the Sun: A Poetical Dialogue by Brother Tommaso Campanella. *Berkeley: University of California Press* 1981: 7–27.
Dougherty, James. 1986. Geometries of Perfection: Atlantis, Utopia and the City of the Sun. *Soundings: An Interdisciplinary Journal* 69 (3): 310–325.
Dyer, James. 2003. *Hillforts of England and Wales*. Princes Risborough: Shire Publications.
Eastwood, Luke. 2012. *The Druid's Primer*. Alresford: John Hunt Publishing.
Foucault, Michel. 1975/1995. *Discipline and Punish: The Birth of the Prison*. Trans. Alan Sheridan. New York/Toronto: Vintage Books.
———. 1978/2009. *Security, Territory, Population: Lectures at the College De France, 1977–78*, ed. Michel Senellart. New York: St Martin's Press.
Frampton, Kennet. 1992. *Modern Architecture: A Critical History*. London: Thames and Hudson.
Freud, Sigmund. 1919/1955. The 'Uncanny'. In *The Standard Edition of the Complete Psychological Works of Sigmund Freud*, ed. and trans. J. Strachey, vol. XVII, 217–252. London: The Hogarth Press.
———. 1923/2010. *The Ego and the Id*. Eastford: Martino Fine Books.
———. 1929/1994. *Civilization and Its Discontents*. Mineola: Dover.
Frugoni, Chiara. 2005. *A Day in a Medieval City*. Chicago/London: University of Chicago Press.
Gandelsonas, Mario. 1998. The City as the Object of Architecture. *Assemblage* 37: 128–144.
Goodchild, B. 1990. Planning and the Modern/Postmodern Debate. *Town Planning Review* 61 (2): 119–137.
Goodman, Dena. 1994. *The Republic of Letters: A Cultural History of the French Enlightenment*. Ithaca: Cornell University Press.
Habermas, Jürgen. 1962/1991. *The Structural Transformation of the Public Sphere: An Inquiry into a Category of Bourgeois Society*. Cambridge: MIT Press.
———. 1981/1984. *The Theory of Communicative Action, Volume 1*. Boston: Beacon Press.

———. 1988/1999. Actions, Speech Acts, Linguistically Mediated Interactions, and the Lifeworld. In *Jürgen Habermas on the Pragmatics of Communication*, ed. Maeve Cooke, 215–256. Cambridge: Polity.

Halbwachs, Maurice. 1950/1992. *On Collective Memory*. Chicago: The University of Chicago Press.

Hansen, Mogens Herman. 1997. The Copenhagen Inventory of Poleis. In *Development of the Polis in Archaic Greece*, ed. Lynette Mitchell and P.J. Rhodes, 5–12. London: Routledge.

Harding, Dennis. 2012. *Iron Age Hillforts in Britain and Beyond*. Oxford: Oxford University Press.

Herodotus of Halicarnassus 440 BCE/1879. *Histories*. Trans. Henry Cary. London: George Bell & Sons.

Hersey, George L. 2000. *Architecture and Geometry in the Age of the Baroque*. London/Chicago: University of Chicago Press.

Howard, Ebenezer. 1898. *To-morrow: A Peaceful Path to Real Reform*. London: Swan Sonnenschein.

Hurwit, Jeffrey M. 1999. *The Athenian Acropolis: History, Mythology, and Archaeology from the Neolithic Era to the Present*. Cambridge: Cambridge University Press.

Husserl, Edmund. 1936/1970. *The Crisis of European Sciences and Transcendental Phenomenology: An Introduction to Phenomenological Philosophy*. Evanston: Northwestern University Press.

Jacobs, Jane. 1961. *The Death and Life of Great American Cities*. New York: Vintage Books.

James, Susan E. 2017. *The Feminine Dynamic in English Art, 1485–1603: Women as Consumers, Patrons, and Painters*. London/New York: Routledge.

Jerman, James, and Anthony Weir. 2000. *Images of Lust: Sexual Carvings on Medieval Churches*. New York: Routledge.

Johnson, David M. 2011. *Socrates and Athens*. Cambridge/New York: Cambridge University Press.

Kelly, Thomas. 1977. *A History of Argos to 500 B. C.* Minneapolis: University of Minnesota Press.

Koch, John T., and Antone Minard. 2012. *The Celts: History, Life and Culture*. Vol. I. Santa Barbara: ABC-CLIO.

Kostof, Spiro. 1991. *The City Shaped: Urban Patterns and Meanings Through History*. Boston/Toronto/London: Little, Brown and Company.

Kuchenbuch, David. 2015. Architecture and Urban Planning as Social Engineering: Selective Transfers Between Germany and Sweden in the 1930s and 1940s. *Journal of Contemporary History* 51 (1): 22–39.

Laërtius, Diogenes. 3 cent CE/1959. *Lives and Opinions of Eminent Philosophers*. Cambridge, MA: Harvard University Press.

Lancaster, H. Carrington. 1945. The Château de Richelieu and Desmaretz's Visionnaires. *Modern Language Notes* 60 (3): 167–172.

Lay, M.G. 1992. *Ways of the World*. Sydney: Primavera Press.

Lefebvre, Henri. 1974/1992. *The Production of Space*. Oxford: Blackwell.

Lewis, Michael J. 2016. *City of Refuge: Separatists and Utopian Town Planning*. Princeton/Oxford: Princeton University Press.

Locke, John. 1689/2005. An Essay Concerning Human Understanding. In *An Essay Concerning Human Understanding in Focus*, ed. Gary Fuller, Robert Stecker, and John P. Wright. London/New York: Routledge.

Macintosh, Alison A., Ron Pinhasi, and Jay T. Stock. 2017. Prehistoric Women's Manual Labor Exceeded That of Athletes Through the First 5500 Years of Farming in Central Europe. *Science Advances* 3(11), Published Online November 29, 2017. https://doi.org/10.1126/sciadv.aao3893.

Marinatos, Nanno. 2010. *Minoan Kingship and the Solar Goddess: A Near Eastern Koine*. Urbana: University of Illinois Press.

More, Thomas, Sir St. 1516/2016. *Utopia*, ed. George M. Logan. Cambridge: Cambridge University Press.

Morgan, Kathryn A. 1998. Designer History: Plato's Atlantis Story and Fourth-Century Ideology. *Journal of Hellenic Studies* 118: 101–118.

Murray, Timothy. 1987. *Theatrical Legitimation: Allegories of Genius in Seventeenth-Century England and France*. Oxford/New York/Toronto: Oxford University Press.

Newman, Oscar. 1996. *Creating Defensible Space*. Washington, DC: U.S. Dept. of Housing and Urban Development, Office of Policy Development and Research.

O'Meara, Dominic J. 2005. *Platonopolis: Platonic Political Philosophy in Late Antiquity*. Oxford: Clarendon Press.

Panofsky, Erwin. 1957. *Gothic Architecture and Scholasticism*. New York: Meridian.

Pappas, Nickolas. 1995. *The Routledge Guidebook to Plato's Republic*. London/New York: Routledge.

Pruthi, Raj. 2004. *Prehistory and Harappan Civilization*. New Delhi: APH Publishing.

Rahmstorf, Lorenz. 2018. Of Middens and Markets: The Phenomenology of the Market Place in the Bronze Age and Beyond. In *Market as Place and Space of Economic Exchange: Perspectives from Archaeology and Anthropology*, ed. Hans Peter Hahn and Geraldine Schmitz, 20–40. Oxford/Philadelphia: Oxbow Books.

Riegl, Alois. 1908/2010. *The Origins of Baroque Art in Rome*, ed. and trans. Andrew Hopkins and Arnold Witte. Los Angeles: Getty Research Institute.

Rosenau, Helen. 1983. *The Ideal City: Its Architectural Evolution in Europe*. London/New York: Methuen.

Salingaros, Nikos A. 1998. A Scientific Basis for Creating Architectural Form. *Journal of Architectural and Planning Research* 15 (4): 283–293.

Slater, T.R. 1999. Geometry and Medieval Town Planning. *Urban Morphology* 3: 107–116.

The Telegraph. 2011. Medieval London: 10 Disgusting Facts, April 5.

Thomas, Carol G., and Craig Conant. 2009. *Citadel to City-State: The Transformation of Greece, 1200-700 B.C.E.* Bloomington/Indianapolis: Indiana University Press.

Thompson, Edward H. 1999. *J.V. Andreae: Christianopolis*. Dordrecht: Kluwer.

Toorians, Lauran. 2008. Endlicher's Glossary, an Attempt to Write Its History. In *Celtic and Other Languages in Ancient Europe*, ed. J.L. García Alonso, 153–184. Salamanca: Ediciones Universidad de Salamanca.

Venturi, Robert. 1966/2008. *Complexity and Contradiction in Architecture*. New York: Museum of Modern Art.

Vico, Giambattista. 1744/1968. Scienza Nuova. In *The New Science of Giambattista Vico*, Revised Translation of the Third Edition of *Scienza Nuova*, ed. Thomas G. Bergin and Max H. Fisch. Ithaca: Cornell University Press.

Wilson, Thomas D. 2016. *The Ashley Cooper Plan: The Founding of Carolina and the Origins of Southern Political Culture*. Chapel Hill: University of North Carolina Press.

Yates, Frances Amelia. 1999. *Giordano Bruno and the Hermetic Tradition*. London: Routledge.

Young, Julian. 1994. *Nietzsche's Philosophy of Art*. Cambridge: Cambridge University Press.

Young, Karl. 1926. The Home of the Easter Play. *Speculum* 1 (1): 71–86.

———. 1930. Dramatic Ceremonies of the Feast of the Purification. *Speculum* 5 (1): 97–102.

CHAPTER 4

From Body Without Organs to City Without Streets

Introduction and Summary

To the extent one can presuppose that the Neolithic round enclosure was used for renewal ceremonies or fertility rituals, later doubling up also as a market area where goods were exchanged, this public space was a precursor to the village and the town square. Due to their solstice ritual use, however, round enclosures seem to have been constructed occasionally at elevated places or on plains where full sunrise and sunset could be observed. Round enclosures, therefore, would not be a physical part of Neolithic settlements which, in contrast, had evolved usually along fresh surface water sources, such as near rivers or natural springs.

Rivers or seashores, however, would often serve also as a crude form of natural sewage disposal as continued to be the case till early modernity and beyond. Up until approximately the time of the modern demographic transition the treatment of effluent in settlements, inclusive of cities, was by way of conveyance of raw sewage to a natural body of water, for example, a river or sea, where it would be diluted and dissipated. Tanners disposed of animal acids, dried blood, fat, hair and other waste residue by unloading them into the river stream. Streets too, not underground sewers, carried wastewater and were the landing grounds of household waste of the widest range. As urban populations past the Middle Ages continued to grow, streets, particularly in the lower sections of cities and towns, endured intensifying defilement.

A. Akkerman, *Philosophical Urbanism*,
https://doi.org/10.1007/978-3-030-29085-6_4

The physical abasement of streetscapes left urban squares intact. From Neolithic round enclosures to the Greek Agora and theater, the Roman forum and the medieval market square, the open public space was a powerful element in the display of power by authorities. It was also the market square that, in time, gave rise to the public sphere of the Enlightenment. Whereas most streets accreted haphazardly through centuries, urban squares were designed, lined with edifices of eminence. Evolution of urban squares, thus, was far from unambiguously affirmative. Serving gatherings of the community, the public space, as a communal open-air area within the geographic boundaries of a settlement, had undergone a contradictory progression over the millennia between the round enclosure of the Neolithic Demographic Transition and the urban square of the Industrial Revolution. On the one hand, the communal open-air space could be said to have been the informal medium conducive to spontaneous advancement of language, art and other forms of human expression, as well as socio-economic interaction, communication and religious discourse.

On the other hand, the late antiquity had marked also the beginning of malediction in designed open urban space, much worse than the physical contamination of streets. Initially, the Neolithic round enclosure and later the town's market square were utilized for popular celebrations, ritual proceedings or spontaneous encounter by people, while the access route leading to the Neolithic round enclosure or later the town's main street was used as a processional avenue, quite possibly well-maintained and perhaps ornate or paved. It was likely with the ascent of the Roman Empire at the turn of the common era that some public stairs, major streets or open areas adjacent to city walls attained a sinister earmark, never witnessed before. At Rome's Esquiline Gate, an important concourse at the entrance to the city, public torture and capital punishment by crucifixion of slaves and real or imagined criminals took place. At the Colosseum, as well as at arenas of other Roman cities, prisoners would fight for their lives with hungry beast to the cheers of the crowd. While tickets needed to be purchased to watch gory gladiatorial fights at the Arena, the spectacle of extreme cruelty exacted upon the condemned at the Esquiline urban precinct was free for all to watch. The purpose was not only to punish and humiliate the offender, but also to intimidate the public and to divert its attention from the authority onto the condemned in a way of ghastly entertainment.

The use of the agora, forum or urban squares as sites of public executions was unheard of. The desecration of designed open-air public space within cities for the purpose of display of torture and execution seems to have emerged with the slow decline of Feudalism in Europe and intensified thereafter. Urban squares became frequent execution sites during the French Revolution. In comparison, rituals performed at the Neolithic round enclosure, the equivalent or precursor of the public square, probably never included executions as a tool of domination over a community.

Even during the darkest days of the Roman Empire, most public squares were exempt from becoming sites of public torture and executions. Such munificence was not conferred on the urban square of the Middle Ages and the Renaissance. Throughout the Middle Ages and beyond, the city's square as a place of markets, discourse and commemoration had attained the additional function as occasional place of murderous amusement and intimidation through public infliction of agonizing physical torment and execution. The French Revolution, inspired by the Enlightenment, was notorious for Parisian squares that became officially sanctioned killing fields where the urban crowd could witness public executions by guillotine. In London until the nineteenth century, schoolchildren during Execution Days watched public hangings of criminals in a festive atmosphere featuring everything from carnival sweets to pickpockets and prostitutes in the crowd.

The abolition of public executions in public space came only after the repeated outcry of handful of intellectual elites, rather than by the government on its own. Charles Dickens and Jeremy Bentham were among those protesting the officially sanctioned disgrace of executions in public space, and the despicable practice was abolished in mid-nineteenth century throughout Europe, while in the United States public executions were permitted into the twentieth century.

In his book, *Discipline and Punish*, Michel Foucault looks at the abolition of public torture and rejects the claim of humanization of penal justice.

> In this transformation, two processes were at work. They did not have quite the same chronology or the same raisons d'être. The first was the disappearance of punishment as a spectacle. The ceremonial of punishment tended to decline; it survived only as a new legal administrative practice. [The second was a] utopia of judicial reticence: take away life, but prevent the patient from feeling it; deprive the prisoner of all rights, but do not inflict pain; impose penalties free of pain. (Foucault 1977: 8–11)

The primary objective of this transformation was not to create a more humane penal system, Foucault points out, but rather, the transformation was simply a stage of a continuing trajectory of subjugation. The application of surveillance in hospitals or prisons, due to Bentham, or to urban planning as proposed by Ledoux, for example, still follows the same trajectory. As a substitute to violence by the overseeing authority, discipline can be attained through surveillance as a nonverbal relationship of intimidation emanating from the authority onto the persons observed (Foucault 1977: 195–230).

Facilitating varying intensity of surveillance turns into an apparatus of power the entire physical structure within which a population is observed: prison inmates, hospital patients or city dwellers. Foucault concludes: "The formation of disciplinary society is connected with a number of broad historical processes – economic, juridico-political and, lastly, scientific – of which it forms a part" (Foucault 1977: 218). On further perusal, the wide scope of such broad historical processes that had led to the modern concept of urban surveillance, as well as its modes of impact, are intertwined with other defining notions of urban modernity: uniformity, automation and superscale.

One of the main reasons for the development of urban surveillance was the progression of informal interpersonal modes of communication, many of which took place in the streets. Throughout much of modernity urban open spaces facilitated not only public expressions of protest against authorities but were also venues of riots and uprisings centered on cities. Authorities saw increasingly streetscape layout as a preventative means to counter disobedience and violence aimed against the ecclesiastical or the secular sovereign authority. Urban surveillance has contributed to the loss of informal gathering places in the city's streets and to the increasingly foreboding apprehension of placelessness in later modernity.

4.1 Demise of Public Space: Modes of Desecration, from Antiquity to Modernity

Jeremy Bentham's furtherance of his Panopticon prison project ought to be seen in the broader context of his lifelong opposition to the death penalty, and his outspoken support for the abolition of slavery and physical punishment (Crimmins 1988). The founder of utilitarianism, Bentham in his design of the Panopticon leaves no doubt as to his belief that surveillance, as a crime prevention mode, is better and more efficient than cruel punishment (Bedau 1983).

Bentham's concept and Ledoux's celebrated ideal city project, Chaux, helped galvanize urban planning across France on the eve of the French Revolution. Surveillance and crowd control became an important consideration in time of intensifying migration into cities, while the ensuing growth of trade and increasing traffic between and within cities had also called for improvement in streetscape infrastructure for public health, safety and security. Already throughout the Renaissance many urban streets transformed from accreted, warped voids between houses and edifices, onto aligned conduits for more efficient traffic, particularly of animal-drawn carts. By the time of the Enlightenment urban planning in France was addressing, similar to Bentham's utilitarianism, crime prevention as an ideological premise. Foucault explains the precepts of urban planning on the eve the French Revolution, taking the project of Pierre Vigné de Vigny (1690–1772) at Nantes as a specific example:

> It involved cutting routes through the town, and streets wide enough to ensure four functions. First hygiene, ventilation, opening up all kinds of pockets where morbid miasmas accumulated in crowded quarters, where dwellings were too densely packed. So, there was a hygienic function. Second, ensuring trade within the town. Third, connecting up this network of streets to external roads in such a way that goods from outside can arrive or be dispatched, but without giving up the requirements of customs control. And finally, an important problem for towns in the eighteenth century was allowing for surveillance, since the suppression of city walls made necessary by economic development meant that one could no longer close towns in the evening or closely supervise daily comings and goings, so that the insecurity of the towns was increased by the influx of the floating population of beggars, vagrants, delinquents, criminals, thieves, murderers, and so on, who might come, as everyone knows, from the country. (Foucault 1978/2007: 33–34)

Foucault example from Nantes points to two trends in the evolution of urban streetscapes, extending millennia from early antiquity to modernity. On the one hand, there had been a trend of gradual physical deterioration in streets, mainly due to insufficient or lacking sewage and stormwater disposal, along with increasing street crime sustained by rapid urban population increase, poverty, inadequate public lighting and correspondingly insufficient policing, all of which urban planning of the Enlightenment was meant to improve. On the other hand, public space in cities had gradually transformed from a medium of community ritual and spontaneous encounter onto an alien geographic space increasingly degraded to formal interaction between strangers, culminating in late twentieth century's urban space that came to be perceived mainly in terms of personal safety and defense. Policing and urban planning were meant to mutually support the joint goal of public safety, surveillance and crowd control, thereby, inadvertently, reducing further informal contact between people.

The first aspect of the historic trend, physical deterioration of urban space, manifested itself mainly in streets that doubled as sewers, struggling to carry away waste from a wide range of activities, including slaughtering and waste disposal. With increasing population in cities, streets in the Middle Ages would become cesspools of effluent. Sewage ran down the streets as cities, most of which had no functioning sewer system, relied on occasional rain showers to wash the wastewater away and drain it as runoff into the local watershed. In medieval European cities, small natural waterways used for carrying off sewage were sometimes covered over, functioning as sewers.

In some cities, even in ancient times, flat stones with a narrow edge were applied to provide even and solid surface to main streets. This practice continued into the Middle Ages when many cities embarked on paving their main streets. Particularly on steeper streets setts, quarried stones shaped into regular form were used to allow surface grip to horses' hooves. On some other streets cobble-sized stones were used for pavement. Most of the medieval urban streetscapes, however, remained unpaved. Thus, for example, Philippe Auguste, King of France (r. 1180–1223), ordered most urban streets to be paved, but centuries later the project remained still unaccomplished. Even on paved streets there were no sidewalks, and some paved streets were concave so that people could walk on the sidelines while sewage was running down the middle of the street.

As opposed to the physical deterioration, the human interaction on Europe's medieval streets was vibrant and often positively stimulating. The weekday usually began with the ringing of the church bell in early morning, announcing the first mass of the day and the end of the night watchman's duty. Street shops opened providing produce, meats or fish before the first meal of the day. Markets opened later in the morning, noisy with merchants shouting out advertising of their wares. Meaningful encounter among townsfolk occurred informally during the weekday's working or shopping action, or at street performances by wandering troupes. Throughout the late afternoon shops began shutting down, heralding the coming diurnal period of rest.

Into early modernity, the second aspect of the historic degradation of urban streetscapes was due to rapid urbanization and population increase, often resulting in social deprivation manifest in crime but even more so, in mutual alienation among urban dwellers. To the average urban dweller most of the rest became strangers. In a letter to his contemporary, Jean-Louis Guez de Balzac, Descartes uses the simile of a city walk to express his joy of anonymity as much as presaging Existentialist alienation from, and contempt for urban crowds:

> I can walk out each day in the bustle of the crowds with as much freedom and ease as you have in your paths, and I pay no more attention to the people I meet than I would to the trees in your woods or the animals that graze there. (Gaukroger 1995: 188)

In Descartes' other brief streetscape description, in Book II of his *Meditations*, surveillance or the gaze yield an air of estrangement as the darker human attributes emerging from early modernity. Descartes had asked a question that in time came to be known for its notoriety: "What do I see from the window beyond hats and cloaks that might cover artificial machines, whose motions might be determined by springs?" (Descartes 1641/1924: 232). The setting of Descartes' question is routinely viewed as a scene whereby "there are men crossing the square in the rain underneath the hats and cloaks which are alone visible from an upper window" (Baker 2000: 625). In his question, Descartes was expressing the estrangement of the individual toward the rest of the urban crowd, conferring a gloomy urban association upon the philosophical quandary regarding the existence of other minds (Matthews 1986).

In parallel to the physical and social aspects of degradation of urban public space ran also physical and social aspects of overt public spectacle originating in the ruling authority. The one kind of a spectacle was glitter and size of edifices, the cathedrals and the palaces, usually on public squares that, as a result and in contrast to most streets, were well-maintained. Whereas the agora, the Roman forum and the town square had been usually well looked after, no commensurate attention was afforded to most streets. The disparity in upkeep between streets and squares had been increasing over time, corresponding to the increasing size and glamor of buildings erected on urban squares.

The other kind of spectacle was yet another aspect of social degradation: the public punishment and humiliation of offenders. It is in this regard that Bentham's rejection of capital and any other corporal punishment is relevant in the context of public space within or adjacent to cities. While human sacrifice or executions took place in archaic societies, the public display of infliction of suffering upon the condemned seems to have been perfected only with the rise of the Roman Empire. Such deviance went hand in hand with the well-documented, progressively declining mental capacity of Roman emperors who instigated and directly or indirectly sanctioned public torture and executions:

> In ancient Rome, death was dictated by social class. At one end, the patricians and the equestrians were allowed to poison themselves in private. At the other, slaves were publicly crucified. Although this form of public execution is now associated with the death of Jesus Christ, it was a common form of capital punishment when Roman Empire was at its height [...] When Spartacus' slave rebellion was crushed in 71 BC, the victorious Roman general Crassus crucified 6,000 slaves along the Appian Way, the main Road leading into Rome from the south. As the Roman general and later emperor Titus was putting down the Jewish revolt and beginning the siege of Jerusalem in AD 70, he was at one point also crucifying 500 Jews a day. Josephus, in his *History of the Jewish Wars*, claimed that so many Jews were crucified outside the city walls 'there was not enough room for the crosses and not enough crosses for the bodies.' [...] The condemned person was made to carry his cross as part of a public procession to the place of crucifixion outside the city walls. It was usually sited on high ground, so that the spectacle could be seen by the maximum number of people, and it would be near a road, to serve as a warning to passers-by [...] In ancient Rome, there were plenty of other ways to meet a public death. In one case 4,500 prisoners were tied to stakes in groups of thirty at the Forum [...] The arena was a popular method of executing condemned criminals, Christians, and other religious or political dissenters perceived to be enemies of the State. (Cawthorne 2006: 9–16)

As antiquity in Europe was giving way to the Middle Ages, the reason for the upkeep of the urban square attained sometimes an ominous streak. The authority's public demonstration of power in the ending of a human life, amplified by the ritual of public torture and execution through to modernity, was performed at squares or in open spaces outside city walls. Extreme violence publicly witnessed was a nonverbal message by the authority communicating warning to any potential violators of law and order. Even when the condemned were common criminals, their public execution was a threatening missive to anyone challenging the authority. Through public viewing of their anguish, usually a part of the capital punishment, the condemned persons served primarily an instrument of power to the overseeing authority.

4.2 Body Without Organs: The Medieval Urban Square as the Theatre of Cruelty

The Tarpeian Rock, a steep cliff about 25 meters high, overlooking the Forum of Ancient Rome, was used during the Roman Republic as an execution site of murderers, traitors, perjurers and rebellious slaves. The Gemonian Stairs, also in full view of the Forum, were used for public display of condemned who were usually first strangled, or otherwise executed elsewhere in the city, then flung down the stairs. Corpses were usually left to rot on the staircase for extended periods of time, scavenged by dogs, then thrown into the Tiber River.

A common physical instrument of public torture and execution through the Middle Ages, though less frequently used afterward, was the breaking wheel or execution wheel. Catherine of Alexandria, now Saint Catherine of the Wheel (c. 287–305) was said to have undergone public torture at the wheel, and then execution, at the age of 18. Also known as Catherine's Wheel or simply the Wheel, it was used to break the condemned person's bones, and was completely abolished only in the early nineteenth century. The practice has been seen a symbolic re-enactment of a previous mode of penalty, where people were literally driven over by a wagon (Spierenburg 1984: 71).

Since 200 CE in western Asia the interest of the authority in public executions was demonstrated by public trampling of the condemned by elephant in public space adjacent to city walls. While enemy soldiers were most commonly found at the receiving end of this unusual punishment, petty criminals found guilty of offenses such as tax evasion were also subject to execution by elephant. This practice lasted well into at least the fourteenth century. The Muslim traveler Ibn Battuta, visiting Delhi in the 1330s, wrote of a public execution: "If the order was to cut him to pieces, the elephant would do so with his irons, and then throw the pieces among the assembled multitude" (Lee 1829: 146–147).

Urban open places were used as a common venue for public torture and executions throughout medieval Europe, even though mostly unrecorded.

In England executions were recorded only since the twelfth century when public hangings at the gallows took place at Tyburn, then just outside London city walls, at a junction of two Roman roads. Another place of public executions in London was Smithfield, a large open space close to the city and a popular place for public gatherings. Along with Tyburn, Smithfield was for centuries the site of public execution of heretics and dissidents in London. The Scottish nobleman Sir William Wallace was executed in 1305 at West Smithfield.

Execution Day ceremonies were designed to reinforce the power of authorities. Between the fourteenth and eighteenth centuries, dozens of people were publicly beheaded on the Tower Hill, outside the walls of the Tower of London, where for two centuries also a permanent scaffold stood.

During the period 1509–1547, reign of Henry VIII of England, thousands of people were executed, some of them publicly on Tower Hill. The author of *Utopia*, Sir Thomas More was beheaded in 1535 at Tower Hill, and so were other nobles, and although many beheadings were not public, the severed heads were sometimes placed on spikes along London Bridge or other public places.

Mary Queen of Scotts was beheaded in front of some hundred spectators in 1587, her severed head displayed from a bay window of Fotheringhay Castle, Northamptonshire, England, so that crowds of people could see that the Scottish queen was dead. Back in London, Lincoln's Inn Fields, the largest public square in the city at the time, laid out in the 1630s, was the site of the public beheading of Lord William Russell in 1683, for the attempt to assassinate King Charles II.

Contrasting the solemnity of public executions of nobility were the gruesome festivities surrounding the execution of common criminals in England:

> Huge crowds gathered along Oxford Road (London's Oxford Street) to see condemned prisoners being taken from Newgate Prison to Tyburn. Popular offenders were showered with flowers and unpopular ones pelted with rotten vegetables or stones. [...] Around the gallows at Tyburn were wooden stands where spectators paid two shillings for a good view. The largest and most desirable stand was Old Mother Proctor's Pews, named after their owner. The whole affair had a carnival feel about it with crowds singing and chanting, and street vendors selling gingerbread, gin and oranges. (Cawthorne 2006: 78–79)

In the sixteenth century the Tyburn style gallows, known as the Tree, were erected in the middle of the road in west London, as a deliberate landmark and very obvious warning to passers-by. Elsewhere in Europe, on Dam Square in front of the Amsterdam town hall public executions were taking place on justice days during the Middle Ages and the Renaissance (Spierenburg 1984: 81–82).

The Gibbet of Montfaucon, outside Paris city walls, was erected in the late thirteenth century to become the main gallows in France, and stood there until the time of Louis XIII in the seventeenth century. Criminals and traitors were publicly executed here by hanging, their dead bodies displayed for all to see. In Paris at the turn of the sixteenth century, many of the public squares were designed under Henri IV, King of France. Henri's purpose was not so much to promote public encounter of exchange of goods and ideas between people, as it was to design spaces for the posturing of might through spectacle of military parades and drills and to display power in front of those who fell under his jurisdiction. Such subtlety was mild in comparison to the places of execution that Parisian public squares were to be used for during the Enlightenment and the French Revolution (Cockburn 1994; Tackett 2015: 312–339).

The Decapitation Bridge in Ghent, Belgium, was located beside the city's castle, public decapitation of murderers and rapists having taken place here till late sixteenth century. In papal Rome executions used to be held publicly in Campo de' Fiori, where, on February 17, 1600, the philosopher Giordano Bruno was burnt alive for heresy. In 1437 at the Old Town Square in Prague Jan Roháč, a peasant leader, and several other rebels, were publicly executed. Also at the Old Town Square public torture and execution of 27 Czech Protestant leaders took place in 1621 by the Austrian Catholic Habsburg authorities. Many public executions during the French Revolution were conducted at Place de la Concorde, one of the major public squares in Paris, and at Place du Trône-Renversé, now Place de la Nation, where victims were guillotined in the open-air.

Less horrifying but no less deplorable was the use of pillory and stocks at urban squares by authorities as a form of spectacular punishment and humiliation of offending individuals, and entertainment mixed with intimidation of the public. A prominent pillory, where the condemned were publicly flogged, was situated next to the statue of King Charles at Charing Cross, London. Typically, a person condemned to the stocks was subjected to a variety of abuses, ranging from having refuse thrown at them, to prickling and whipping of the unprotected limbs. Pillory or public stocks as a means of humiliation of offenders at public spaces, and intimidation of the public, were used till the nineteenth century in various places throughout Europe. Area around the pillory was a popular place of street entertainment, the surrounding taverns bursting with patrons watching the crowds and the humiliation of the malefactors.

In 1849, Charles Dickens, along with 30,000 other spectators, watched the hanging of the Mannings, a notorious pair of murderers. Hastening the Capital Punishment Amendment Act 1868 was a letter Dickens sent to *The Times* in 1849, in which he argued against public executions:

> [...] The horrors of the gibbet and of the crime which brought the wretched murderers to it, faded in my mind before the atrocious bearing, looks and language, of the assembled spectators. When I came upon the scene at midnight, the shrillness of the cries and howls that were raised from time to time, denoting that they came from a concourse of boys and girls already assembled in the best places, made my blood run cold. As the night went on, screeching and laughing, and yelling in strong chorus of parodies [...] were added to these. When the day dawned, thieves, low prostitutes, ruffians and vagabonds of every kind, flocked on to the ground, with every variety of offensive and foul behaviour.[...] I am solemnly convinced that nothing that ingenuity could devise to be done in this city [...] could work such ruin as one public execution, and I stand astounded and appalled by the wickedness it exhibits. (Dickens 2011: 201)

The intention to intimidate the public through the display of executions clearly backfired on the authority. Public execution sites brought together the worst among the mob, and the carnival atmosphere at the sight of gallows only made light of yet another killing, albeit one sanctioned by the courts. Within such context the Bentham's notion of the Panopticon applied in the fashion of Ledoux's urban design seems a relatively progressive crime prevention alternative to public punishment.

The last public beheading in England took place outside the Newgate Prison in 1820. That was after a period that lasted likely more than a millennium, throughout which the penalty for men convicted of criminal offenses or high treason, graduated during the reign of King Henry III (1216–1272) to being hanged, drawn or quartered at urban squares, in front of crowds that often took part in the executions. A convicted criminal or traitor, after being emasculated and disemboweled, was then beheaded. The criminal's body without the organs was then chopped into four pieces, all in public view. Centuries later, *Body without Organs* was the ghoulishly absurd notion that were to haunt the schizophrenic playwright Antonin Artaud (1947/1976: 571), only to return, by way of the philosophers Gilles Deleuze and Simone de Beauvoir, onto a metaphor for urban space of twentieth century's modernity.

On May 29, 1868, the Capital Punishment Amendment Act received Royal Assent, thus ending public executions in the United Kingdom as of that date. From then on all prisoners sentenced to death were to be executed within the prison walls, their bodies buried in the prison grounds. A decade later, in Paris, Henry Becque, a playwright, introduced his idea of the Theatre of Cruelty. Following WWI, Antonin Artaud (1896–1948) reintroduced the Theatre of Cruelty as a "sense of violent rigor and extreme condensation of scenic elements that is the cruelty on which it is based [...]" (Artaud 1968: 66). The Capital Punishment Amendment Act 1868 and Artaud's Theatre of Cruelty seem to mark a course of changing attitudes to the public display of cruelty, but also to venues where these attitudes have been expressed, from late antiquity to modernity.

4.3 Evolution of the Urban Streetscape: Rise of Uniformity and the Loss of Place

It was due to their multiple functionality, often dubious or deplorable, that public open spaces such as urban squares were relatively well-maintained throughout the Middle Ages and modernity. In contrast to such communal open spaces in settlements from prehistory to early modernity, urban *streets* have endured gradual deterioration, largely due to increasing population and the lack of correspondingly adequate sewerage. One purpose of orthogonal design of streetscapes in Bronze Age Harappan cities, in Iron Age Greek colonies by Hippodamus of Miletus, and in Roman encampments and outposts, was also to facilitate water and effluent runoff from streets, sometimes leading straight into the sea, rivers or deep ravines. The great success of irrigation canals of the Indus Valley civilization, and increasing prosperity of their agricultural settlements, may have inspired the linear streetscapes of cities such as Harappa and Mohenjo-daro. The perceived aesthetic impact of orthogonal streetscapes, however, was probably secondary.

Aesthetic impact of streetscapes was also not the primary concern in Roman colonies, where orthogonal street layout had religious and possibly overriding consideration of visual oversight and surveillance. The facility

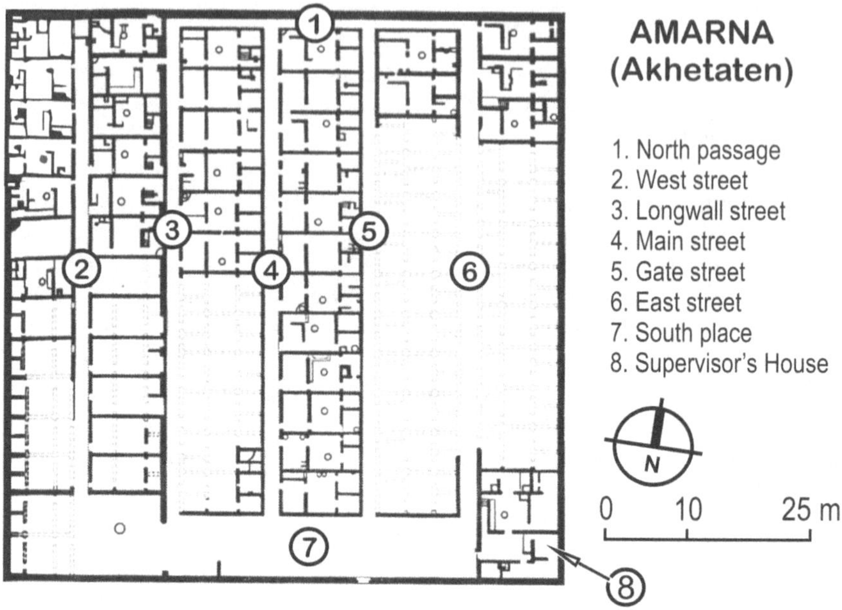

Fig. 4.1 Workers colony, el-Lahun, c. 2000 BCE. (Drawn by Paul Van Pul)

of surveillance and control as an inherent urban feature could be seen implicit already in the plan of an Egyptian workers' colony at el-Lahun, c. 2000 BCE (Rawlinson 1887: 343) (Fig. 4.1).

Most ancient and medieval towns, however, were slowly accreting in a nonlinear fashion over centuries. The vast majority of medieval cities and towns of the Romanesque and the Gothic formed around market squares or at major crossings of trading routes such as roads or rivers. Cities grew on both banks of rivers which were used for washing and as sewage and animal waste runoff. While cities and towns usually expanded in a centrifugal fashion, often through annexation of surrounding villages, a few medieval towns grew in parallel with, and along trading routes. Streetscape patterns that ensued from such physical conditions, though not entirely random, were further constrained by a physical city boundary of bulwarks or walls.

Pending affordability of each community, medieval towns built various earthworks around them for protection, and affluent cities had walls with gates locked at night or at a time of crisis such as during a siege. Stores and workshops doubled as homes for the tradesman that owned them and worked in them. Since most of the townsfolk were illiterate each shop owner used a picturesque relief carved or affixed on a building, showing the trade the shop occupying it was engaged in. Lined with houses upon which such amiable signage was carved, until early modernity most urban streets had no official names and houses were not even numbered. The population of an entire city did not exceed a few thousand people and was often the size of a present-day neighborhood. Only capital cities such as Baghdad or Constantinople had populations approaching a million inhabitants during the Middle Ages. Other major medieval populations were about 400,000 in Beijing, 350,000 in Cairo, 200,000 in Delhi, 200,000 in Milan, 200,000 in Antwerp, 100,000–300,000 in Paris, or 75,000 in London. In most urban communities, much smaller than their capital counterparts, people knew each other, and personal interaction on streetscape was often one of spontaneity.

The tangled layout of streetscapes, which for brevity could be referred to as Romanesque, was common to most cities throughout much of Europe, the Mediterranean and beyond, during antiquity and the Middle Ages. The capriciously interlaced Romanesque streetscape, while functioning quite efficiently at low speed of traffic and mostly low-rise buildings, in a way also complemented the nature outside the city. The narrow, twisted streets needed to accommodate only people and animal traffic, rather than greenery, while a short walk outside the city there always was nature in abundance. It was the city wall, or some other earthworks, as physical growth boundary, that had forced the tortuous urban streetscapes of antiquity and the Middle Ages, while at the same time giving rise to a compact city-form. But growing and restive population in the twisting streets of medieval cities became a menacing concern to the ruling authority, the local lord or the sovereign.

To facilitate easier oversight of authorities, cities were divided administratively into parishes and districts. In response to the fear of insurgency, and in contrast to the actual physical layout of streetscapes in most ancient and medieval cities, the evolving conception of the Ideal City, since antiquity, had been nourished by the notion of harmonious and uniform geometry offering the implicit ability of surveillance, and seemingly affording correspondingly harmonious, hierarchically and rigidly stratified community. Such geometric and social uniformity emerges from Plato's physical layout of both Atlantis and Magnesia. In Magnesia Plato instructs to

> divide the city into twelve portions, first founding temples to Hestia, to Zeus and to Athene, in a spot which we will call the Acropolis, and surround with a circular wall, making the division of the entire city and country radiate from this point. The twelve portions shall be equalized by the provision that those that are of good land shall be smaller, while those of inferior quality shall be larger. (Laws V: 745c–811ce)

Neo-Platonic schemes of an ideal city were multiple visionary plans from Platonopolis of Plotinus in the early Middle Ages to the Garden City of Ebenezer Howard at the turn of the twentieth century. The one type of actually constructed settlements that followed a Neo-Platonic scheme, already in the Middle Ages, were the *bastides*, small fortress towns, mostly in southern France and northern Spain. Contrasting the prevailing random street plan the *bastides* were orthogonally planned *new towns*, during the late twelfth and early thirteenth century. Expedited by the twelfth-century translation into Latin of Euclid's *Elements* by the English monk Adelard of Bath (1080–1152), the *bastides* were laid out on a grid plan with a central square, drawing also on the standard plan of the Roman military settlement, the *castrum*, a millennium earlier (Boerefijn 2000; Lilley 1998). By the fourteenth century, at the onset of the Gothic, there were hundreds of *bastides* throughout south-western Europe (Morris 1994: 119–120), inspiring the construction of *new towns* of the later Middle Ages and the Renaissance.

At late Middle Ages, the architectural and urban aspects of the Gothic coincided with the slow inception of urbanization in north-western Europe, in particular. Contrasting the simplicity of the *bastide* plan was the elaborate vertical feature of the built form of the Gothic design, facilitated by the advent of the pointed arch. Ribbed vaults and transverse arches had allowed for ever taller vertical construction, primarily of churches, that had spread from regions of the Islamic Mediterranean, in the south, and from the northern Holy Roman Empire, throughout much of today's Germany. The Gothic streetscape largely absorbed the earlier built Romanesque random street pattern. But planned new towns, often at the behest of Feudal overlords, proliferated, usually laid out on orthogonal grid patterns of varying sophistication such as the Florentine *new towns* of San Giovanni and Terra Nuova (Friedman 1988: 120–129). *New towns* of the Renaissance were often laid upon more elaborate geometric plans, either orthogonal (such as Freudenstadt) or radial-concentric (such as Henrichemont) (Fig. 4.2).

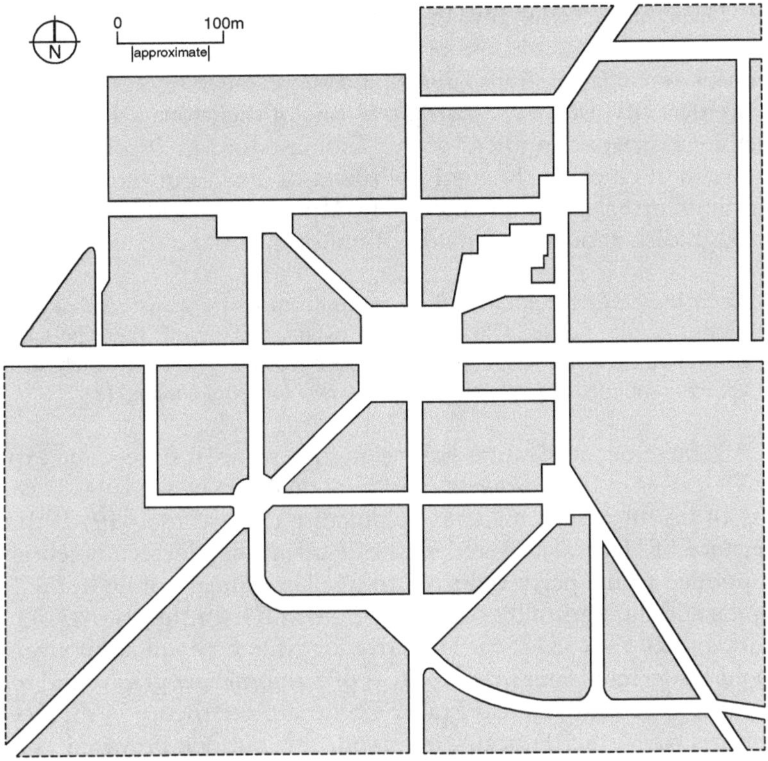

Fig. 4.2 Henrichemont, central France, c. 1609. (Drawn by Paul Van Pul)

The contrast between the rigorous, mostly orthogonal, streetscape of the *bastides* and the new towns, against the tortuous, fickle streetscapes of the Romanesque gave rise to René Descartes' reverie in the stove-heated room of autumn 1619. From Descartes' description of his multifold dream emerges subtle impact that streetscape contrast carries upon consciousness. The archaic Romanesque city-form and its tortuous narrow streets juxtaposed against the new towns' orderly aligned streetscapes from which the Renaissance planned urban environment had arisen, had given Descartes a pivotal perceptual contrariety.

To the defenders of ancient cities, as Aristotle notes in *Politics*, Book VII, it was the chaotic streetscape layout that had warranted safety against outside intruders. And since no two random streetscapes could be identical, the Dionysian deems any streetscape unique. The Platonic, or Neo-Platonic streetscape, on the other hand, embraces the Apollonian urban outlook of elegance, transparency and surveillance suggesting a template of an ideal city plan: "He that knows one of their towns, knows them all, they are so like one another," writes Thomas More in *Utopia* (1516/2016) of the ideal cities on the mythical island of Utopia in the early sixteenth century. Further praise is conferred by More upon the ideal cities of Utopia through early modernity's marks of uniformity, size and automation:

> Their buildings are good, and are so uniform that a whole side of a street looks like one house. The streets are twenty feet broad; there lie gardens behind all their houses.[...] Their doors have all two leaves, which, as they are easily opened, so they shut of their own accord. (*Utopia* II)

All the same, admiration for the aligned planned streetscape expressed by Descartes in the *Discourse on Method* ought to be seen also in the context of his interest in military architecture (Descartes 1619/1991). The elegance of Haussmann's redesign of Paris in the nineteenth century corresponded to the perceived need to quell continual rioting in the French capital and the fears of his patron, Napoleon III, for the survival of his rule (Kirkland 2013: 233–256). The drive for symmetry and harmony in space could be detected later in some areas of economic geography and regional science, such as Walter Christaller's central place theory in the twentieth century. As Michel Foucault observed, this outlook prevailed as an academic abstraction through much of the industrial age:

> [A] good country is one that, in short, must have the form of a circle, and the capital must be right at the center of the circle. A capital at the end of an elongated and irregular territory would not be able to exercise all its necessary functions. In fact, this is where the second, aesthetic and symbolic, relationship between the capital and the territory appears. The capital must be the ornament of the territory. But this must also be a political relationship in that the decrees and laws must be implanted in the territory [in such a way] that no tiny corner of the realm escapes this general network of the sovereign's orders and laws. The capital must also have a moral role, and diffuse throughout the territory all that is necessary to command people with regard to their conduct and ways of doing things. The capital must give the example of good morals. The capital must be the place where the holy orators are the best and are best heard, and it must also be the site of academies, since they must give birth to the sciences and truth that is to be disseminated

> in the rest of the country. Finally, there is an economic role: the capital must be the site of luxury so that it is a point of attraction for products coming from other countries, and at the same time, through trade, it must be the distribution point of manufactured articles and products, etcetera. (Foucault 1978/2009: 27–28)

By the eighteenth century the death knell was sounded for medieval streetscapes. The pressure of demographic growth led to the demolition of ramparts and the incorporation of suburbs, with new districts built around a more rational grid layout. This involved the creation of new urban squares, street lighting and wider roadways. Royal orders in France and papal orders in the late sixteenth and early seventeenth century restricted or prohibited vaulted construction while enforcing alignment rules designed to widen roadways. Prohibited in many cities through France and Italy were ledges or other projections jutting forth from the facades of buildings. An eighteenth-century regulation in Paris forbade construction of streets narrower than 9.75 meters.

Worth noting is the kind of street encounter among inhabitants associated with the Platonic, Apollonian urban disposition, and the one emanating from the Dionysian, incorporated in Aristotle's urbanist outlook. Where the Apollonian urban character of solemnity, elegance and transparency implies expectedness in street interaction among inhabitants, the Dionysian, haphazard streetscape foments chance encounter among people passing through it. Whereas Dionysus could be said to imply spontaneity and playfulness on the street, the sun-god Apollo in his patronage of the planned city runs a darker streak through its translucent streetscapes: rigor and disciplined uniformity in design call for restraint behavior, reticence and control. From the Dionysian spontaneity in urban space, a place emerges, slowly, unexpectedly, sometimes in an entirely unforeseen manner. The unacknowledged but overriding human feature of the carefully planned Apollonian layout is the gaze: its transformation in a masterplan is surveillance and the demise of place.

4.4 City Without Streets and the Superscale: De Beauvoir Versus Deleuze

Ever since the rise of modernity, urban planning, most vividly represented by Georges-Eugène Haussmann's redesign of central Paris in the nineteenth century, has embraced, silently within its comprehensive plans, the concept of the ideal city. Rejecting the tortuous, constricted streets of the medieval city-form as counterproductive to transportation, public health and safety and security, Haussmann ratified the rejection of randomness and enigma inherent in spontaneously accreted streetscapes (Chapman 1953). The ensuing comprehensive notion of the masterplan has been typified by an underlying, universal concern for transparency and predictability in streetscapes, primarily as a facility to allow for continual traffic flows, and as prevention of transportation conflicts (DeJean 2014: 122–144).

Urban transportation throughout the nineteenth century and across the North Atlantic regions of Europe and America had been quickly mechanized. The omnibus was introduced as a horse-drawn carriage with the capacity of about a dozen seats, starting its first operation at Manchester, England, in 1824, in 1826 at Nantes, and in 1828 in Paris. Alphonse Loubat (1799–1866), a French inventor helped develop the first horse-drawn tramway at New York City in 1827. In 1832 Loubat built the first tram line, for horse trams, which operated between 1853 and 1855, the year of the World Fair in Paris. In 1860 George Francis Train, an American entrepreneur and inventor, arrived in England, subsequently founding horse tramway companies in Birkenhead and London, and a short-lived horse tramway in Cork, Ireland.

Michaux's invention of the velocipede in 1857 was a serious competition to the horse-drawn traffic in Paris. The success of the velocipede was conditioned by Haussmann's wide, elegant thoroughfares whereby the modern streetscape of Paris gained, in return, an added feature. Supplementing streetscape uniformity and superscale of new tenement buildings had been now also speed. The horse-drawn public traffic could not possibly compete in velocity with the human-powered automation and the traffic speed of the velocipede.

Compressed-air traction cars and steam engines had addressed both speed and mass transit in the second half of the nineteenth century, contributing to the bankruptcy of Michaux's company.

Steam locomotives enclosed in a wooden box structure, in the likeness of a railroad passenger coach, were used in urban transportation in the United States from the 1930s till c. 1865, the end of the Civil War. In England, at Loughborough, Henry Hughes began manufacturing urban steam-tram engines in 1876. Kitson & Co. started to build steam-trams in 1878. On the Continent Georg Krauss converted some horse-drawn trams into steam-trams in Munich and Vienna in 1876, steam-trams were built and operated in cities across France through the 1880s, and a steam-tram engine was introduced at Brno, then Habsburg Empire, in 1884. Throughout the 1890s steam-trams in cities begun to be replaced by electricity-powered trams. Streets and sidewalks became also important components of budding metropolitan transportation since walking and public transit came to be inseparably connected.

Urban transportation became a major consideration in the management and administration of cities at the turn of the twentieth century. With the advent of the internal combustion engine and the mass production of automobiles urban transportation problems throughout the industrialized west accrued and intensified. Geographic expansion of cities and incessant construction of urban freeways fed into one another, requiring comprehensive plans to address myriad urban issues emerging as a result, all the while leaving street corners and sidewalks increasingly neglected. Whereas in the medieval city sidewalks were nonexistent, in the industrial and postindustrial metropolis the commitment to alleviate traffic congestion and to ensure safety, bound with the need to share the public thoroughfare, had gradually led to streets being split into different sections, a simple walk between them now constricted by rights of way or crosswalks. The street, now cluttered with signposts of information, instructions, advertisements and directives became a stage for mechanized choreography directed, first by traffic police, later by traffic lights. To a considerable extent, automation, uniformity and superscale, on the one hand, and the masterplan, as a latter-day version of the Ideal City on the other hand, fed into one another while consideration of human scale in streetscapes became secondary or nonexistent (Jacobs 1961: 178–186).

Seeds of the disconnection between human and metropolitan scales, however, were planted long ago with the archetypal notion of the ideal city. Countering these trends were the attempts of William Morris and Camillo Sitte, defending traditionalist attitudes to urban planning during modernity. Writing in Vienna at the end of the nineteenth century Sitte campaigned against uniform alignment of streets and urban squares, hailing the erratic streetscapes of early Renaissance cities in northern Italy. In his *City Planning according to Artistic Principles* Sitte pointed to humanistic values of spontaneity, serendipity and surprise associated with irregular streetscapes (Collins and Collins 2006: 61–63; Sitte 1889/2006: 170–190). In a careful survey of late-medieval and Renaissance towns in northern Italy, Sitte had observed that instead of superficial green nature, the medieval streetscape design—or often, the entire lack of planned street layout—have created a built environment tantamount to extension of the natural environment. The unpredictable streetscapes of the medieval town had been impregnated with a visual mastery of small plazas, facilitating continuous human encounter as well as prevalence of visual appeal.

Sitte's groundbreaking analysis of medieval squares has been extended later in the twentieth century in a typology of urban squares by the Luxembourg architect Rob Krier (1975/1979: 30–50). Significance of human scale in urban planning and design to counter uniformity and automation has been emphasized by the likes of William 'Holly' Whyte (1980) or Marshall McLuhan (1966). Toward late modernity authors have singled out human scale as defying culpable urban features of contemporary metropolis through the affirmation of human encounter in city-form (Lund 2003; Wilson 2017).

But it was also during the twentieth century that two of the most monumental Neo-Platonic urban ideals have been carried out as artificial cities, planted on virgin grounds. The masculine, Apollonian myth of the ideal city was symbolized by Walter Burley Griffin in his award-winning radiocentric design proposal for Canberra in 1912 (Proudfoot 1994). And in Brasília, the federal capital city of Brazil, now a UNESCO World Heritage Site, the same masculine myth had been perpetuated during the 1960s (Palazzo and Saboia 2012).

The Harvard economist, Edward Glaeser, has recently dubbed Brasília a "warning to urban dreamers" (Glaeser 2012). Brasília's monumental public buildings, the work of the architect Oscar Niemeyer, are breathtaking in their magnificence: Catedral Metropolitana, Congresso Nacional, Praça dos Trés Poderes, Palácio da Justiça or Palácio do Planalto. But to Simone de Beauvoir Brasília's overwhelming abundance of superscale in its buildings is only matched by its emaciated urban voids functioning as passage circuits, transit circulations between origins and destinations. To de Beauvoir Brasília is a city without streets:

> [W]hat possible interest could there be in wandering about among the six- or eight-story quadra and super-quadra, raised on stilts and all, despite superficial variation, exuding the same air of elegant monotony? [...] But the street, that meeting ground of riverside dwellers and passers-by, of stores and houses, of vehicles and pedestrians – thanks to capricious, always unexpectedly changing mixture – the street, as fascinating in Chicago as in Rome, in London as in Peking, in Bahia as in Rio, sometimes deserted and dreaming, but alive even in its silence, the street does not exist in Brasilia [...] (De Beauvoir 1992: 273)

The predicament of Brasília as city without streets accords, in the Americas as elsewhere, with many planned urban environments (Dunnett 2000). De Beauvoir's unflattering view of Brasília's streets presents it as an amalgam of elegant shipment conduits representing the planned city of modernity and postmodernity. A view that the city is inherently nothing, *but* a conglomerate of transit channels and junctions has emerged in *A Thousand Plateaus* (1980/2004) by Giles Deleuze and the psychiatrist Pierre-Félix Guattari. The essence of *any* city since antiquity, the authors seem to argue, is a compendium of transit lines and intersection points. Urban quintessence, to Deleuze and Guattari, arises from the abstraction of city-form as a universally common feature of origins, destinations and the transport of people and goods between them. Such globalizing abstraction to Deleuze is 'deterritorialization,' a theoretical platitude applicable to any city, anytime and everywhere. Deleuze's deterritorialized notion of the city could be read as a description of urban essence, yet it could also be easily interpreted as a facetious critique suggesting that the entirety of the urban range since antiquity has been dehumanizing:

> [T]he town is the correlate of the road. The town exists only as a function of circulation, and of circuits; it is a remarkable point on the circuits that create it, and which it creates. It is defined by entries and exits; something must enter it and exit from it. It imposes a frequency. It effects a polarization of matter. Inert, living or human; it causes the *phylum*, the flow, to pass through specific places, along horizontal lines. It is a phenomenon of *transconsistency*, a *network*, because it is fundamentally in contact with other towns. It represents a threshold of deterritorialization, because whatever the material involved, it must be deterritorialized enough to enter the network, to submit to the polarization, to follow the circuit of urban and road recoding. The maximum deterritorialization appears in the tendency of maritime and commercial towns to separate off from the backcountry, from the countryside (Athens, Carthage, Venice). (Deleuze and Guattari 1980/2004: 477–8)

To Deleuze the city, since antiquity, has been a deterritorialized system, a capsule of transit hubs. Human condition in the deterritorialized city is derived from urban voids of a built environment that processes people. Such urban voids are not meant to facilitate human encounter, but "Deleuze sees in them a vague beginning of an impersonal ethics of urban modernity" (Saldanha 2017: 155).

Far from being solely a postmodern urban circumstance, Deleuze's deterritorialized urban environment, runs a streak reaching, indeed, to antiquity. Human automatism and uniformity, as features of the social structure of Plato's ideal cities, Atlantis and Magnesia, correspond to the ideal city's physical layout, and run in parallel, from the legislators down to common folk. As if uniformity and automatism were inherent, archaic features, going millennia back to Plato's ideal cities:

> The conditions suppose a population with no disrelish for [...] social regulations, who will tolerate life-long limitation of property, restrictions such as those we have proposed on procreation, and deprivation of gold and other things which it is certain, from what has been said already, that the legislator will prohibit; they presuppose further the central position of the capital, and the distribution of the dwelling-houses over the territory, as he has prescribed, almost as though he were telling his dreams or fashioning a city and its inhabitants out of waxwork. (*Laws* V: 746 in Hamilton and Cairns 1962: 1330)

The Platonic and Neo-Platonic urban outlooks through the ideal cities of late antiquity, Middle Ages, to modernity, and to Deleuze's postmodern abstraction of the city, could hardly be typified other than Apollonian. Yet Apollonian transparency arising from the designed streetscape of an urban plan finds its place also in Aristotelian urbanism, where it matches equally important Dionysian randomness emanating from the enigmatic streetscape of haphazard accretion. Where the Apollonian outlook of the Platonic ideal city is a uniform shell of solemnity, Aristotelian urbanism embraces both the Apollonian and the Dionysian in city-form.

Deleuze's deterritorialization, as an abstracting feature held in common by all, detached from the uniqueness of the particular, is a generic observation upon the environment and does not apply to cities only. In *Anti-Oedipus* Deleuze and Guattari introduce the deterritorialized notion of Body without Organs, or BwO. A template to any human body, BwO is perceived as a shell, a masterplan, into which each and every body evolves to be fitted:

> The body without organs is an egg: it is crisscrossed with axes and thresholds, with latitudes and longitudes and geodesic lines, traversed by gradients marking the transitions and the becomings, the destinations of the subject developing along these particular vectors. (Deleuze and Guattari 1972/1977: 19)

Here Deleuze's definition of BwO is thought of as the preconceived, generalized or deterritorialized notion of a body. De Beauvoir condemnation of Brasília, too, views the city without streets, CwS, as a pre-planned conglomerate of beautiful superstructures and fashionable transport channels in lieu of ordinary streets. Humans are inserted into the CwS only after it has been first conceived of, and planned, then constructed, ostensibly in their behalf. This is also how North American suburban subdivisions have been built.

In contrast, the exact urban opposite to CwS is the Romanesque town that had been accreting slowly, with no preconceived centralized plan but through multitude of ad-hoc actions by residents who were an integral part of their town from its very inception. The closest a North American city precinct has come to such a traditionalist urbanist notion, was the poverty-stricken inner-city neighborhood. During the twentieth century such precincts had dwindled, not due to prosperity of the incumbent residents, but often due to invasive gentrification of affluent suburbanites seeking urban authenticity, nowhere to be found in their own suburban neighborhoods. Twentieth century's invasive gentrification of the inner city had often proceeded along the lines reminiscent of de Beauvoir's CwS.

Initially, Deleuze and Guattari had presented the deterritorialization concept in a definition of BwO, as an absurdity image of the residual *after* a body leaves a particular, geographically delineated, space. On this initial definition, BwO is the body's lingering leftover of sounds, smell and fluidity,

> a new dimension of the schizophrenic body, an organism without parts which operates entirely by insufflation, respiration, evaporation and fluid transmission (the superior body or body without organs of Antonin Artaud). (Deleuze 1969/1990: 100–101)

Thinking away the organs, the body's sounds, smell and fluidity become the essence of BwO. Similarly CwS too could be perceived as the residual of noise, smell and fluidity left behind by the metropolis of modernity and postmodernity. Within CwS resentment becomes the human condition. The author Milan Kundera, Deleuze's French compatriot of Czech descent, describes how noise, smell and speed emanating from contemporaneous streetscapes yield nothing but the disgusting and the gruesome of the city, much as BwO in Deleuze:

> She said to herself: when the onslaught of ugliness became completely unbearable, she would go to the florist and buy a forget-me-not, a single forget-me-not, a slender stalk with miniature blue flowers. She would go

> out into the streets holding the flower before her eyes, staring at it tenaciously so as to see only that single beautiful blue point, to see it as the last thing she wanted to preserve for herself from a world she had ceased to love [...] Suddenly, the sharp sound of a motorcycle pierced her being. She could not help but immediately look towards something that had caused such physical pain: a young girl in jeans, her black hair waving behind her, erect on a small motorcycle as if she were sitting at a typewriter; it had no muffler and made a horrific noise. [...] Agnes looked at the hair streaming behind that noisy aggressor and realized she intensely wished the death of that girl. [...] Her hate immediately frightened her and she said to herself: the world has arrived to the frontier of something disastrous; if it crosses it, everything will turn to madness: the people will wander through the streets with forget-me-nots in their hands or will kill each other on sight. It will take very little, the drop of water that overflows the glass: just one car, person or decibel more. (Kundera 1990: 21–22)

Kundera shows the condition of human beings in the metropolis where what used to be a street has transformed through modernity onto a battlezone of mutual estrangement. Against the street as a medium of affirmative human encounter in antiquity, the Middle Ages, and early modernity stands Kundera's passionate profiling of pernicious relations between people emerging from the metropolis past modernity. The waxwork paradigm of uniformity and automatism in the *Laws*, an ingrained feature of the ideal city from Plato to Le Corbusier, meets its own byproduct: human alienation.

4.5 Benjamin and Lefebvre on Power Images in the Metropolis

In 1869, a few years after James Hobrecht was fired by the Royal-Prussian urban planning police, the *Panopticum* opened its doors in Berlin, Prussia's capital. Venture of the brothers Louis and Gustave Castan, it was Germany's very first gallery of wax figurines, exceedingly popular at its beginning. By the end of the nineteenth century some 500 three-dimensional life-size figurines of the famous and the powerful were on display, among them the poet and scientist Johann Wolfgang von Goethe, the geographer Alexander von Humboldt, the composer Richard Wagner or the philosophers Friedrich Schiller and Immanuel Kant. But it was here too that an ordinary commoner could gaze at the entire Prussian royal family, and scrutinize from close-up some of Prussia's most prominent politicians and military leaders.

A few steps from the street under the watchful eye of the *Baupolizei*, the door into the *Panopticum* opened to a dreamlike experience of whimsical capabilities. Here a pedestrian from a moment ago could become a preponderant sovereign on a fantasy-stage of make-believe power acquired with the purchase of an entrance ticket. In the dream house of the *Panopticum*, the roles became inverted and the powerful came under the surveillance of the commoners.

To Benjamin, the Berlin waxworks were a shrine to the compulsion for surveillance, whereby everybody in the metropolis now could scrutinize even the most powerful into minute detail:

> The universalism of the nineteenth century has its monument in the waxworks. Panopticon: not only does one see everything, but one sees it in all ways. (Benjamin 1933/1999: 531)

The wax museum allowed for a temporary, albeit fake, switching of the surveillance roles between authority and its subjects, thereby channeling the subjects' resentment onto a harmless venue of mendacious elation.

Yet resentment has often remained the human response to the urban masterplan, the modern rendition of the Ideal City. To be fitted onto spatial template of a preconceived built environment, and processed through its streetscapes and edifices, has been the tacit presumption of the master designer—baron Haussmann, the *Baupolizei*—projected upon urban inhabitants. Oblivious to communal memory or history the master designer has conjured disaffection through the masterplan, evoking the authority-subject fold as a gentle inflection of the master-slave relationship.

The modern version of the master-slave composite was introduced by Georg Friedrich Hegel in the *Phenomenology of Spirit* (1807/1998), tracking the evolution of human consciousness. In the initial stage human individual's consciousness recognizes the Other, as a reflection of self. Each of two interacting individuals seeks recognition of self through struggle between one another. The result is the division of individuals into two classes. One class of individuals sees freedom as the most important value without which life is not worth living. These individuals become the Masters. The other class of individuals sees life as more important than freedom and, accordingly, is willing to sacrifice freedom for life. These individuals then become the Slaves laboring for their Masters. Gradually, however, Slaves become masters of their own labor, freeing themselves from their paradigmatic Masters.

Expanding Hegel's master-slave relationship onto architectural construction Daniel Purdy points to Benjamin, showing that in addition to the sovereign or the designer who initiate a building, the Masters, and the human subjects producing the construction, the proletariat, or the proverbial Slaves, there is a third component, the building's occupants:

> Benjamin adapts the master/slave dialectic Hegel develops in his theorization of [the Tower of] Babel, so the construction and office workers are the real benefactors [...], for they develop a communal consciousness as laborers quite opposed to the isolation of self-aggrandizing monarchs. [...] Benjamin implicitly follows Hegel's master/slave dialectic when he famously states in his "Work of Art" essay that buildings are either perceived visually by tourists, who see with the detached aesthetics gaze of kings and princes, or touched bodily by those who live and work within them. (Purdy 2011: 6–9)

Twentieth century's failure of grand housing projects displacing inhabitants of poverty-stricken, run-down neighborhoods in the inner city, into uniform high-rise towers, is a fitting example to the loss of memory in a masterplan. Largely oblivious to communal history and relations in the inner-city neighborhoods, planners and architects of patronizing urban renewal projects, such as the most infamous Pruitt-Igoe (Heathcott 2012), seemed to have inadvertently echoed the absolutist docket evoked by master-slave. Urban estrangement and the urban masterplan are in an unholy embrace of a process of mutual feedback and reinforcement yielding alienation and resentment.

In antiquity Aristotle introduced the notion of the master-slave relation as one of two basic social features of the archaic city-state, and of the primeval household—the social building block of the city-state. Aristotle's other relation within the household and the city-state is an alternative to the Apollonian master-slave relation, and it is the alliance of the sexes:

> The first coupling together of persons [...] to which necessity gives rise is that between those who are unable to exist without one another: for instance the union of female and male for the continuance of the species (and this not of deliberate purpose, but with man as with the other animals and with plants there is a natural instinct to desire to leave behind one another being of the same sort as oneself). (*Politics* VII: 11)

In contrast to the Apollonian master-slave relation, or the authority-subject fold, the alliance of the sexes could be characterized as primarily

Dionysian. If so, there is a parallel between the physical makeup of the city's streetscape advocated by Aristotle and the city's social makeup as observed by him in actuality. The streetscape mix endorsed by Aristotle, consisting of elegant alignment as well as haphazard randomness, seems to reflect the Apollonian-Dionysian social formation of the city.

Whereas antiquity and the Middle Ages saw on occasion the Aristotelian streetscape mix, often more by default rather than by a deliberate plan, modernity could be characterized by incessant attempts to expunge the Dionysian from its urban environments. Henri Lefebvre has shown the imbalance in modernity's unconscious gender projection upon the built environment, producing overt urban spectacles, on the one hand, and hidden places of contempt, on the other hand:

> [T]he existence within space of *phallic verticality*, which has a long history but which at present is becoming more prevalent, cries out for explanation. The same might be said apropos of the general fact that walls, enclosures and facades serve to define both a *scene* (where something takes place) and an *obscene* area to which everything that cannot or may not happen on the scene is relegated: whatever is inadmissible, be it malefic or forbidden, thus has its own hidden space on the near or the far side of a frontier. [T]he city, has an underground and repressed life, and hence an 'unconscious' of its own. (Lefebvre 1974/1991: 36)

To be fitted onto spatial template of a preconceived built environment, and processed through its streetscapes and edifices, is the tacit presumption of the master designer, administered to urban inhabitants. In the master-slave relationship as defined by Hegel (1807/1998: 116–119), the master has a need to be acknowledged as such by the slave, and this acknowledgment is integral to its role as a dominating authority. Lefebvre argues that similar acknowledgment, along with uniformity, is sought through the masculine production of urban space in order to preserve the authority-subject embrace:

> The arrogant verticality of skyscrapers, and especially of public and state buildings, introduces a phallic or more precisely a phallocratic element into the visual realm; the purpose of this display, of this need to impress, is to convey an impression of authority to each spectator. Verticality and great height have ever been the spatial expression of potentially violent power. [I]t posits, presupposes and imposes homogeneity in the subdivision of space. (Lefebvre 1974/1991: 98)

Two millennia of urban planning might be viewed as a gradual process of authoritarian masculine projection, culminating in urban modernity, as observed by Lefebvre. Urban planning throughout history has remained largely oblivious to the Aristotelian urbanist notion. The way this lapse has come to a head through modernity is demonstrated by the conversion of places of contempt in the city, onto places of deceptive misrepresentation. Poor precincts from early modernity that somehow survived the wrath of economic supply and demand have become outdoor museums of urban past extinguished, or places of abode to upscale consumers of urban development, rather than casual features of authentic urban space. Earlier in the twentieth century the Aristotelian premise of integrated streetscapes had been decidedly dislodged by the myth of the superscale, summed up by heroic slogans, such as "Make No Small Plans" by Daniel Burnham, or "Growth and Expansion," the adage of Fred "Big Daddy" Gardiner, the mayor of Metro Toronto.

Contrasting the uniform and automated superscale approach has been the call for small, open-air streetscapes, injected into the mainstream environment of the metropolis to purposely provide deliberate ambiguity, and to bring contrast to, and relief from the incessant attempt to rationalize all urban space. A Neo-Romanesque streetscape of tumultuous lanes and twisting narrow alleyways inserts a spatial message of enigma, non-reason, unpredictability and surprise into the automated metropolitan superscale. A reconciliation of the metropolis with the Neo-Romanesque is also a synthesis of Aristotelian streetscapes with Lefebvrian critique of urban uniformity and authoritarian superscale. The purpose of efficient movement between an origin and a destination in a homogeneous urban space becomes secondary in a streetscape that provides a deliberate counterbalance to urban uniformity, of which the North American metropolis is a prime example.

The tangle of downtown back-alleys and surface parking lots, presently the blight of many a North American city, could be instrumental as a Neo-Romanesque response heeding the Aristotelean call that "a city should adopt both plans." Conversion of existing leftover spaces, often-time places of contempt in the city, into small streetscapes, interjected within the mainstream metropolitan space, has been a silently emerging stratagem speaking to the urban crisis of postmodernity through revival of the traditional urban form (Akkerman and Cornfeld 2010; Pemberton and Phillimore 2018). Addressing the Aristotelian riddle of antiquity as well as the Lefebvrian critique of modernity and postmodernity is the Neo-Romanesque urbanist response that could instill human scale into the streetscapes of the metropolis.

Bibliography

Akkerman, Abraham, and Ariela Cornfeld. 2010. Greening as an Urban Design Metaphor: Looking for the City's Soul in Leftover Spaces. *The Structurist* 49/50: 30–35.

Allen, Reginald E. 1985. *Greek Philosophy: Thales to Aristotle*. New York: Free Press.

Artaud, Antonin. 1947/1976. "To Have Done with the Judgment of God," a Radio Play (1947). In *Antonin Artaud: Selected Writings*, ed. Susan Sontag, 555–574. Berkeley: University of California Press.

———. 1968. The Theatre of Cruelty. In *The Theory of the Modern Stage*, ed. Eric Bentley, 55–75. New York: Penguin.

Baker, Gordon. 2000. The Senses as Witnesses. In *Descartes' Natural Philosophy*, ed. Stephen Gaukroger, John Schuster, and John Sutton, 604–629. New York: Routledge.

Bedau, Hugo Adam. 1983. Bentham's Utilitarian Critique of the Death Penalty. *The Journal of Criminal Law and Criminology* 74 (3): 1033–1065.

Benjamin, Walter. 1933/1999. *The Arcades Project*. Trans. Howard Eiland and Kevin McLaughlin. Cambridge, MA: Belknap Press.

Boerefijn, W. 2000. Geometry and Medieval Town Planning: A Contribution to the Discussion. *Urban Morphology* 4 (1): 25–28.

Cawthorne, Nigel. 2006. *Public Executions: From Ancient Rome to the Present Day*. London: Arcturus Publishing.

Chapman, Brian. 1953. Baron Haussmann and the Planning of Paris. *Town Planning Review* 24 (3): 177–192.

Cockburn, J.S. 1994. Punishment and Brutalization in the English Enlightenment. *Law and History Review* 12 (1): 155–179.

Collins, George R., and Christiane Crasemann Collins. 2006. *Camillo Sitte: The Birth of Modern City Planning*. Mineola: Dover.

Crimmins, J.E. 1988. Hatchet for Paley's Net: Bentham on Capital Punishment and Judicial Discretion. *Canadian Journal of Law and Jurisprudence* 1 (1): 63–73.

de Beauvoir, Simone. 1992. *Hard Times: Force of Circumstance – The Autobiography of Simone de Beauvoir*. Vol. 2. New York: Paragon.

DeJean, Joan. 2014. *How Paris Became Paris: The Invention of the Modern City*. New York/London: Bloomsbury.

Deleuze, Gilles. 1969/1990. *The Logic of Sense*. New York: Columbia University Press.

Deleuze, Gilles, and Pierre-Félix Guattari. 1972/1977. *Capitalism and Schizophrenia – Volume I: Anti-Oedipus*. Toronto: Penguin Group Canada.

———. 1980/2004. *Capitalism and Schizophrenia – Volume II: A Thousand Plateaus*. London/New York: Continuum.

Descartes, René. 1619/1991. Letter to Beeckman, 24 January 1619. In *The Philosophical Writings of Descartes: Volume 3, the Correspondence*, ed. John Cottingham, Robert Stoothoff, Dugald Murdoch, and Anthony Kenny, 1. Cambridge/New York: Cambridge University Press.

———. 1641/1924. Meditations. In *The Method, Meditations and Philosophy of Descartes*, ed. John Veitch, 206–282. New York: Tudor Publishing.

Dickens, Charles. 2011. Letters of Charles Dickens: 1833–1870. In *Georgina Hogarth and Mary Dickens*. Cambridge/New York: Cambridge University Press.

Dunnett, James. 2000. Le Corbusier and the City Without Streets. In *Modern City Revisited*, ed. Thomas Deckker, 56–79. London/New York: Spon Press.

Foucault, Michel. 1977. *Discipline and Punish: The Birth of the Prison*. New York: Vintage Books.

———. 1978/2007. *Security, Territory, Population: Lectures at the College De France, 1977–78*, ed. Michel Senellart. New York: Palgrave Macmillan.

Friedman, David. 1988. *Florentine New Towns: Urban Design in the Late. Middle Ages*. Cambridge, MA: MIT for the Architectural History Foundation.

Gaukroger, Stephen. 1995. *Descartes: An Intellectual Biography*. Oxford/New York: Oxford University Press.

Glaeser, Edward. 2012. Brasília Is a Warning to Urban Dreamers. *Financial Times*, December 9.

Hamilton, Edith, and Huntington Cairns, eds. 1962. *The Collected Dialogues of Plato*. Princeton: Princeton University Press.

Heathcott, Joseph. 2012. Planning Note: Pruitt-Igoe and the Critique of Public Housing. *Journal of the American Planning Association* 78 (4): 450–451.

Hegel, Georg Wilhelm Friedrich. 1807/1998. *Phenomenology of Spirit*. Delhi: Motilal Banarsidass Publishers.

Howard, Ebenezer. 1898. *To-morrow: A Peaceful Path to Real Reform*, Second Edition Printed in 1902 as *Garden Cities of To-Morrow*. London: Swan Sonnenschein.

Jacobs, Jane. 1961. *Death and Life of Great American Cities*. New York: Random House.

Kirkland, Stephanie. 2013. *Paris Reborn: Napoléon III, Baron Haussmann, and the Quest to Build a Modern City*. New York: St. Martin's Press.

Krier, Rob. 1975/1979. *Urban Space*. New York: Random House.

Kundera, Milan. 1990. *Immortality*. New York: Grove Weidenfeld.

Lee, Samuel. 1829. *The Travels of Ibn Battūta*. Holborn: Howell and Stewart.

Lefebvre, Henri. 1974/1991. *The Production of Space*. Oxford: Blackwell.

Lilley, Keith D. 1998. Taking Measures Across the Medieval Landscape: Aspects of Urban Design Before the Renaissance. *Urban Morphology* 2: 82–92.

Lund, Hollie. 2003. Testing the Claims of New Urbanism: Local Access, Pedestrian Travel, and Neighboring Behaviors. *Journal of the American Planning Association* 69 (4): 414–429.

Matthews, Gareth B. 1986. Descartes and the Problem of Other Minds. In *Essays on Descartes' Meditations*, ed. Amélie O. Rorty, 141–152. Berkeley/Los Angeles: University of California Press.

McLuhan, Marshall. 1966. *Understanding Media: The Extensions of Man*. London: Routledge & Kegan Paul.

More, Thomas, Sir St. 1516/2016. *Utopia*, ed. George M. Logan. English translation by Robert M. Adams. Cambridge: Cambridge University Press.

Morris, A.E.J. 1994. *History of Urban Form Before the Industrial Revolution*. London/New York: Routledge.

Palazzo, Pedro, and Luciana Saboia. 2012. Capital in a Void: Modernist Myths of Brasilia. *Traditionalist Dwellings and Settlement Review* 24 (1): 15.

Pemberton, Simon, and Jenny Phillimore. 2018. Migrant Place-Making in Super-Diverse Neighbourhoods: Moving Beyond Ethno-National Approaches. *Urban Studies* 55 (4): 733–750.

Proudfoot, Peter. 1994. *The Secret Plan of Canberra*. Kensington: University of New South Wales Press.

Purdy, Daniel. 2011. *On the Ruins of Babel: Architectural Metaphor in German Thought*. Ithaca: Cornell University Press.

Rawlinson, George. 1887. *Ancient Egypt*. New York: G. P. Putnam's Sons.

Saldanha, Arun. 2017. *Space After Deleuze*. London: Bloomsbury Academic.

Sitte, Camillo. 1889/2006. City Planning According to Artistic Principles. In *Camillo Sitte: The Birth of Modern City Planning*, ed. George R. Collins and Christiane Crasemann Collins, 129–332. Mineola: Dover.

Spierenburg, Pieter C. 1984. *The Spectacle of Suffering: Executions and the Evolution of Repression, from a Preindustrial Metropolis to the European Experience*. Cambridge: Cambridge University Press.

Tackett, Timothy. 2015. *The Coming of the Terror in the French Revolution*. Cambridge: Belknap.

Whyte, William Hollingsworth. 1980. *The Social Life of Small Urban Spaces*. New York: Project for Public Spaces.

Wilson, Helen F. 2017. On Geography and Encounter: Bodies, Borders, and Difference. *Progress in Human Geography* 41 (4): 451–471.

CHAPTER 5

LIA and the Iron Age Cold Epoch: Similitudes and Sequels

Introduction and Summary
The parables of the Garden and the Citadel have a primordial source. The myths of the Garden and the Citadel ensue, respectively, from prior paradigms of the Eternal Return and the *Axis mundi*, the two founding archetypes of environmental gender projection. Human-built environments between the Neolithic and late modernity have retained underlying features that point to archetypal substance in the built environment and the humans within it, over the last ten millennia. The eviron/*mental* configuration of humans is set by the myths of the *Axis mundi* and the Eternal Return, yielding the archetypal ingredients of the Garden and the Citadel. The coevolution and the continuing interaction of the two archetypal prototypes have been molding the history of the built environment and the urban circumstance to this day.

On this view, gender parables seem to have evolved from the primordial paradigm of the Eternal Return onto myths of the Garden or the *Ouroboros*, and from similarly primeval paradigm of *Axis mundi* onto the masculine myth of the Citadel and the Ideal City. Human-built environments between the Neolithic and early modernity have retained underlying features that point to such archetypal substance in the built environment and the humans within it. Throughout the Neolithic and the Bronze Age the feminine Garden myth had been instrumental in the deliberate design of open spaces such as the round enclosures, later the agora, the forum and

A. Akkerman, *Philosophical Urbanism*,
https://doi.org/10.1007/978-3-030-29085-6_5

the *poleis* that grew around them during antiquity. The masculine myth of the Citadel had been similarly projected onto the landscape in the construction of the fort, the hillfort and the fortress town during antiquity. During the Middle Ages features of market towns and fortress towns had often merged, the Garden and the Citadel having continued to evolve and to be projected onto the built environment. In time, however, in later medieval towns and city-states, under a sovereign or an overlord exerting power, often through intimidation, the masculine began to override the feminine in the built environment.

The domination of the masculine over the feminine in city-form had continued to sharpen into modernity. The masculine imposition upon late city-form had marked the formerly feminine features of open spaces in the city by regimentation expressed through automation, superscale and uniformity, the masculine features of the metropolis. Urban spaces, as communal places of joint memory and social transaction have been marginalized or disappeared from the cityscape entirely. The loss of place in the city has occurred usually due to loss of communal memory associated with displacement, imposition of automation, uniformity or superscale upon it. The loss of place has resulted in the diminution of the paradigmatic Garden whereby gardens have usually become the manicured possessions of regimented greenery adjacent to homes, while public parks have often turned into urban danger zones to avoid. Urban squares, imprints of the Garden, have gradually transformed into traffic intersections, and the archetype of the Eternal Return has found an outlet in the automated diurnal repetition ritual of urban commuting. The starry nightly sky, disappearing entirely due to city's light pollution, has deprived the growing child from the perceptual link between the earth and the universe. The archetype of *Axis mundi* found its outlet in the images in the modern transmutation of the Citadel onto dense multitude of urban towers rather than in imagination uniting the earth and the sky.

Loss of place is the eclipse of precincts where urban development has introduced infills of new edifices or transport routes, disconnecting the spatial medium of previously unadulterated human encounter and communal memory. The embattled, vanishing place, quashed by the urban volume of the superscale, however, has remained a feminine facet of the city: it cannot be eliminated from city-form. The feminine of the city always re-emerges elsewhere in the urban space, reborn as a niche of scorn, perhaps a space of urban development of years past, which has failed, turning into disorderly urban jumble. The architect Ignasi de Solà-Morales perceives such urban space as an ever-present *terrain vague*, a placeless urban substance, that can appear inadvertently and always, in niches anywhere throughout the city. Whereas designed volumes and regimented voids define the overt face of city-form, placeless niches in the city are often ignored or treated with contempt by the mainstream community. Authors of the city's masterplan—the modern notion of the ideal city—have often labeled such placeless niches, 'urban decay.' To Jean-Paul Sartre in his philosophical novel, *Nausea*, the urban decay of the *terrain vague*, the contemptuous placeless niche, is the *other*, feminine face of the city. In their multiple emergence throughout the city, placeless niches, *terrains vagues*, disdainful in the eyes of the mainstream urban community, also constitute the urban subconscious. In contrast to the loss of place, placelessness is the unformed urban milieu prior to regimented city-form being forced upon it: it is always the primordial, feminine substance of the city, its subconscious.

Much as Solà-Morales and Sartre see placelessness as the primordial urban space, a parallel can be sought for the structured time of the city. The structured time of the metropolis is defined by calendars and clocks. But such structured, mechanistic time does not apply in the primordial space of the *terrain vague*. Iterations and structured frameworks for repetitive, automatic performance of tasks in the city, against randomness and surprise, point to two different concepts of time traceable to archaic age. The archetype of the Eternal Return could be seen projected by humans through urban automation, as the masculine feature of the city, while the sporadic and inadvertent rebirth of the placeless urban niche throughout the urban space is the feminine impulse of the Eternal Return in the city. In Hindu religion cyclicality of the universe is a key concept. And in Egyptian religion the deity Atum impregnates self, giving birth to the twins Shu and Tefnut, who become two different concepts of time. "Shu is Eternal Recurrence and Tefnut is Eternal sameness" (Allen 1988: 21).

Nietzsche and Benjamin have addressed the Myth of the Eternal Return entirely contrarily. Whereas Nietzsche had absorbed it in its orthodox sense, Benjamin had rejected it as a superstition of bygone age. But beyond the contradictory attitudes it has attracted, the archetype of the Eternal Return had been instrumental not only in myth but also in science. Eternal Return is the mental template in the initial postulation of precession of the equinoxes, put forward by Hipparchus of Nicaea in the second century BCE. During the nineteenth century a hypothesis of cyclical climate change across millennia, based on multiple aspects of changes in the

Earth's orbital and axial features, had been advanced, and later on definitely formulated by Milutin Milankovich in 1941. Such hypotheses could hardly arise from empirical data only, without the mental template of intuition on recurring cycles. Similarly too, a recent hypothesis of periodic extinctions in cycles of 26 million years, proposed by Michael Rampino and Ken Caldeira (2015), could not be drawn up without the same mental template of endless iterations.

Whereas the Eternal Return appears to have a feminine flair, transmuting into the myths of the fecund Garden and the oval *Ouroboros*, *Axis mundi* insinuates mainly masculinity. The archetype of *Axis mundi* had been for millennia the source of the notion of the center, particularly in the physical construction of buildings and cities. As an underlying mythical notion, *Axis mundi* had been conducive in the near-simultaneity in the advent of Iron Age science, technology and patriarchal religion from China to the Mediterranean, during half-a-millennium event which Karl Jaspers labeled Axial Age. Civilizations of the northern hemisphere during this time had been forged by the archetypal myths of the Eternal Return and *Axis mundi*, both of which ought to be ascribed to identical sky patterns across the northern nightly sky.

The significant technological advances in the northern hemisphere during Axial Age, overlapping the Iron Age Cold Epoch (IACE), were largely based on the multiple independent applications and uses of the wheel. Observations of the rotating sky along with consistently bad weather were jointly conducive to both technological inventiveness as well as to novel religious beliefs, all occurring almost simultaneously during the Axial Age. Human discernment of identical sky patterns, particularly in relation to the North Star, throughout the northern hemisphere, was thus also the link between the rise of patriarchal religion and the foundation of science.

Consistent with such an outlook seems to be the broader context of the rise of Axial Age during the Iron Age Cold Epoch and of the Scientific Revolution two millennia later, during the second half of the Little Ice Age (LIA). Similar to Axial Age, patterns in the unfolding of simultaneous discoveries of coordinate geometry, along with the advent of Cartesian rationalism and British empiricism as cornerstones of the Enlightenment and the Scientific Revolution, coincide with newly emergent built environment at the turn of yet another half-a-millennium period, the Little Ice Age (c. 1350–1850 CE). The juxtaposition of IACE with LIA is of interest not only due to the advances made by humanity during these two periods. The IACE was followed by a millennium of climate fluctuations of the Roman Warm Period, the Dark Ages Cold Period and the Medieval Warm Period. Modernity, roughly coinciding with the LIA, had come to a close during the twentieth century, presently followed by climate change believed by many to constitute global warming, possibly also given to considerable fluctuations. Although the impact of current climate change is almost unanimously viewed as grave for the entire planet, the psychocultural sequels of modernity and the LIA upon the present or the future are unknown. But if the millennium-long historic sequel following IACE and Axial Age, marked by strife and upheaval in Europe, is any clue, then the archetypal belief in the Eternal Return could be one of the more profound myths of our civilization.

5.1 Placelessness as a Condition of Modernity: *Nausea* of Jean-Paul Sartre

In a somewhat inclusive sense, a *place* suggests a small geographic niche that stands for the informal, non-institutional stage upon which some communal mutuality and reciprocity has been occurring over a length of time. A streetscape as a meaningful space is forged by dispersal of *places*, small geographic niches such as street corners, alcoves or shops that had attained consummate appeal for people in the vicinity. *Places* so understood could hardly emerge due to a premeditated plan but, more likely, they are the incremental, accreted effect of multiple and repeated human effort and various, indeterminate modes of interaction. Streets that do not entail a *place* are humanly meaningless, and a city that is a mere network of streets without places, is city without streets (or CwS), in the narrative of Simone de Beauvoir. In its absurdity, repugnance and monstrosity, city without streets, is not entirely different from *Body without Organs*.

To an individual, however, the notion of a *place* is more elusive and fleeting, possibly impossible to define. Already a hundred years before de Beauvoir's CwS, Friedrich Nietzsche had pondered what a *place* in the city is. Nietzsche expressed desire for an urban place of reflection, one he could not find in the contemporary city, but one a city should possess nevertheless in some undisclosed, desirable future. In Chapter 4 of his *Gay Science* Nietzsche writes about his beloved Turin:

> It gives me a melancholy happiness to live in the midst of this confusion of streets [...] An insight is needed (and that probably very soon) as to what is specially lacking in our great cities namely, quiet, spacious, and widely

> extended places for reflection, places with long lofty colonnades for bad weather [...] for us godless ones to be able to think our thoughts in them. We want to have ourselves translated into stone and plant, we want to go for a walk in ourselves when we wander in these halls and gardens. (Nietzsche 1882/1974: 224–225)

About the same time, writing in Paris while the renovation scheme of Haussmann was devouring the ancient streetscapes of its historic center, Charles Baudelaire had discerned *place* in the city as an urban niche that *was* but is no longer a geographic space of the past, one that he can recall only in memory. Baudelaire was horrified by the newly carved streetscapes and the ensuing destruction of the run-down medieval places that he had called a home. His longing for the very real past of Parisian streetscapes and their humanity would play a major role in his poetry. In a section entitled, 'Parisian Scenes,' including 18 urban poems within the volume, *The Flowers of Evil*, Baudelaire had launched bitter criticism of urban modernity transpiring in his cherished city, where estrangement was now the prevailing sentiment.

Baudelaire censures the sanitized and geometrically laid out streets of Haussmann's Paris which had displaced the city's villains and anonymous anti-heroes, sources of inspiration to his poems in *The Flowers of Evil*: the beggars ('To a red-haired beggar girl'), the blind ('The Blind'), the prostitute ('The Swan'), the old, or the gambler, in the poem 'The Generous Gambler,' in Baudelaire's other collection, *Paris Spleen* (1869). It is in 'The Swan' where Baudelaire expresses his loathing of Haussmann's geometrically aligned, predictable streetscapes. He sees Paris transformed into a monotonous construct of new, shiny, almost identical edifices that litter the city through urban niches each of which now an alien metropolitan space, and in Baudelaire's memory, a *place* that was, but is no longer (Chambers 1993: 24–29, 221–227; Gasarian 1997).

While *place* to Nietzsche is forged by the longing for a wishful environment in some undetermined urban future, nonexistent at present, to Baudelaire *place* is in the very real, spatially defined but vanished past, a spatial niche *no longer* existent at present. In a strange way, for both Nietzsche and Baudelaire the *place* becomes eloquent and animated through the memory of the past, or through an imagined future, while nonexistent at a present time.

A hundred years before Nietzsche and Baudelaire, arriving in Paris from Turin was the young Genevan philosopher Jean-Jacques Rousseau.

In 1742 when Rousseau entered Paris, the streets and the buildings would have been left largely intact since the Middle Ages, more than a century before Haussmann's urban renewal scheme. Rousseau was then 30, about the same age as Baudelaire when he wrote his *Flowers of Evil*, and only a few years younger than Nietzsche writing his *Gay Science*. The vanished places for which Baudelaire so despaired, were all there, for anyone to see, to walk through and to experience. To today's reader, however, Rousseau's response to Paris was quite striking:

> How much did Paris disappoint the idea I had formed of it! The exterior decorations I had seen at Turin, the beauty of the streets, the symmetry and regularity of the houses, contributed to this disappointment, since I concluded that Paris must be infinitely superior. I had figured to myself a splendid city, beautiful as large, of the most commanding aspect, whose streets were ranges of magnificent palaces, composed of marble and gold. On entering the faubourg St. Marceau, I saw nothing but dirty stinking streets, filthy black houses, an air of slovenliness and poverty, beggars, carters, butchers, criers of diet-drink and old hats. This struck me so forcibly, that all I have since seen of real magnificence in Paris could never erase this first impression, which has ever given me a particular disgust to residing in that capital. (Rousseau 1782/1861: 127–128)

Faubourg St. Marceau, the Parisian suburb that gave the young Rousseau the first impression of the city, was earlier on one of the destinations of tanners and dyers that King Louis XIV expelled from Paris in 1673 due to the sights, pollution and smell of the carcasses of animals used in the manufacture of skins. The tanneries were in full operation at St. Marceau when Rousseau entered it, with the river Bièvre the dumping site of all residuals from the workshops. It is hard to imagine a different response than that Rousseau had expressed, even if it were Baudelaire in his stead. Clearly, Rousseau's disappointment was magnified by his preceding fantasy images of Paris as an ideal city. There are, thus, two urban portraitures at play: the authentic place and the Ideal City. It seems Baudelaire is longing for the authentic place of the past while encountering the ideal city of Haussmann, and Rousseau is longing for the Ideal City he imagined while abhorring the odious experience of the authentic place. There appears to be no clear-cut resolution to the tension between the authenticity of place and the Ideal City.

Similar is the effect later urban renewal had on deprived urban niches, vermin-infested, overrun by crime and poverty-stricken, but brimming with authenticity. Exemplary is the observation of Franz Kafka, twentieth century's foremost existential writer. In 1893 the Law of Sanitation was decreed for Prague's Josefov district, Kafka's childhood residence. Inspired by Haussmann's renewal scheme in Paris, the inner-city precinct of Prague's Josefov was razed and rebuilt anew, meeting a fate similar to central Paris a few years earlier. Ancient houses and crooked lanes were demolished, replaced by fashionable buildings and ordered streetscapes. A stylish arterial road, named none other than Paris Boulevard, cut through the impoverished precinct. No longer would there be litter and raw sewage marking the streets. But here is what Kafka had to say after the sanitized urban renewal of Josefov:

> The dark corners, the mysterious passages, the boarded-up windows, the dirty yards, the noisy beer-shops and the shuttered inns still live in us. We walk through the broad streets of the newly-built town. Yet our steps and our glance are unsure. Innerly we still shiver as we did in the old streets of misery. Our hearts still know nothing of the re-sanitation that has been carried out. The sick old Jewish Town is much more real to us than the new hygienic town now surrounding us. (Frynta 1960: 59–60)

Similar to Baudelaire, to Kafka too the authenticity of a place becomes acute upon its demise, when the production of controlled space expunges the authenticity of a streetscape through the masterplan. Where Lefebvre stopped short of addressing the evolutionary genealogy of production of space, Kafka had given it poignant attention. But it was Christian Norberg-Schulz who conferred upon urban space not only philosophical reflection, but also a historic and psychoanalytic context. It is this context that has led to twentieth century's coining of the notions of placelessness by Ed Relph (1976) and loss of place by Norberg-Schulz:

> We only recognize the fact that man is an integral part of the environment, and that it can only lead to human alienation and environmental disruption if he forgets that. To belong to a place means to have an existential foothold, in a concrete everyday sense. When God said to Adam, 'You shall be a fugitive and a wanderer on the Earth,' he put man in front of his most basic problem: to cross the threshold and regain the lost place. [sic] (Norberg-Schulz 1980: 23)

Lefebvre's observation of homogenous division of urban space speaks directly to the notion of loss of place by Norberg-Schulz, as an inherent failure of twentieth century's metropolis (Stanek 2011: 136). But the loss of place and placelessness are two different notions of urban space. As an ingrained feature of modernity, placelessness enjoins alienation as a primordial urban quality, an initial condition from which both modern humanity and city-form emerge. The loss of place, on the other hand, is the forfeiture of a spatial niche in the city, that was unique to a community and was defined by communal memory.

But how does one distinguish between the loss of place and placelessness in the materiality of urban space? Ignasi de Solà-Morales, the late Spanish architect, has coined the term, *terrain vague*, to which the notion of placelessness corresponds. To Solà-Morales abandoned, obsolete and unproductive geographic niches, ambiguous and obscure, that somehow were wrestled out of procedural land-use controls, in their incertitude convey the primal essence of the city:

> Empty, abandoned space in which a series of occurrences have taken place [...] I will denote by the French expression *terrain vague*. [It] assumes the status of fascination, the most solvent sign with which to indicate what cities are and what our experience of them is. [...] When architecture and urban design project their desire onto a vacant space, a *terrain vague*, they seem incapable of doing anything other than introducing violent transformations, changing estrangement into citizenship, and striving at all costs to dissolve the uncontaminated magic of the obsolete into the realism of efficacy. (Solà-Morales 1995/2014)

To Solà-Morales alienation, a sentiment of estrangement from others, dread and desolation, is a primordial condition, whether expressed in human attitude and feeling, or in the emergence and sustenance of *terrain vague*. Perhaps unwittingly, Solà-Morales confers a feminine trait to *terrain vague*, and to its juxtaposition against the planned structure. His depiction of this confrontation can be interpreted as a struggle between the Apollonian, masculine volume and the Dionysian, feminine void. This is not only a description of contemporary urban configuration but also a call for the recognition and legitimation of the feminine within city-form:

> Today, intervention in the existing city, in its residual spaces, in its folded interstices can no longer be either comfortable or efficacious in the manner postulated by the modern movement's efficient model of the enlightened tradition. How can architecture act in the *terrain vague* without becoming an aggressive instrument of power and abstract reason? Undoubtedly, through attention to continuity: not the continuity of the planned, efficient,

> and legitimized city, but of the flows, the energies, the rhythms established by the passing of time and the loss of limits [... W]e should treat the residual city with a contradictory complicity that will not shatter the elements that maintain its continuity in time and space. (Ibid)

Terrain Vague, as an authentic urban niche, is not a static solid, but an unadulterated recess, one often overlooked initially by the masterplan, and if at all noticed, it is usually treated as a place of contempt. It is a niche of abandon, which has evolved as a result of myriad minute jounces originating in its immediate urban vicinity rather than through a premeditated action. Once discerned, the authority confers on *terrain vague* the derogatory label 'urban decay' with a view of transforming it into a potentially purposeful, utilitarian space, aligned with the mainstream of the city. To Jean-Paul Sartre, the urban decay of *terrain vague*, its placelessness, is the only authentic face of the city, it is its alien, primordial and pre-rational essence, a geographic space that, in his novel, *Nausea*, exudes the primality of the Garden:

> I am on the curb of the Rue *Paradis*, beside the last lamp-post. The asphalt ribbon breaks off sharply. Darkness and mud are on the other side of the street. I cross the Rue *Paradis*. I put my right foot in a puddle of water, my sock is soaked through; my walk begins. [...] I am cold, my ears hurt; they must be all red. But I no longer feel myself; I am won by the purity surrounding me; nothing is alive, the wind whistles, the straight lines flee in the night. The Boulevard Noir does not have the indecent look of bourgeois streets, offering their regrets to the passers-by. No-one has bothered to adorn it: it is simply the reverse side. [...] The town has forgotten it. Sometimes a great mud-coloured truck thunders across it at top speed. No one even commits any murders there; want of assassins and victims. [...] The Nausea has stayed down here, in the yellow light. (Sartre 1938/1964: 24–26)

Henri Lefebvre has alluded to niches of contempt in the city as constituting the city's feminine face. Lefebvre discerns the city's places of contempt as the essence of the otherness against the urban space surrounding them. In one breath he also answers in the affirmative Freud's question, whether the city is a psychical entity (Lefebvre 1974/1991: 36; Section 4.5). The 'reverse side' of the town, the otherness of urban space and the contradiction of the mainstream are also the gist of the assenting notion of Heterotopia introduced by Michel Foucault (1967/1986). Far from being contemptuous, Foucault confers on otherness in geographic space the feminine vigor of the Garden as a sacred place:

> We must not forget that in the Orient the garden, an astonishing creation that is now a thousand years old, had very deep and seemingly superimposed meanings. The traditional garden of the Persians was a sacred space that was supposed to bring together inside its rectangle four parts representing the four parts of the world, with a space still more sacred than the others that were like an umbilicus, the navel of the world at its center (the basin and water fountain were there); and all the vegetation of the garden was supposed to come together in this space, in this sort of microcosm. [...] The garden is the smallest parcel of the world and then it is the totality of the world. The garden has been a sort of happy, universalizing heterotopia since the beginnings of antiquity [...]. (Foucault 1967/1986)

5.2 Automation and Benjamin's *Capitalism as Religion*: Urban Space in Modernity's Rituals

In contrast to fossilized emblems of spatialized, solemn Form of the ideal city, urban decay is the spontaneous, inadvertent and alien expression of haphazard urban authenticity. To the mainstream the place of urban decay is quite meaningless or appalling, to the authority a target of eradication. But niches of urban decay never disappear for good. They may be patched over, erased or eliminated, only to burst open elsewhere in the city, or to slowly emerge from cracks and crannies in other streetscapes. Within contemporary city-form urban decay is a persevering, ever-present, ongoing veracity: the unplanned place, or the unintended aftermath of a plan. It is a spontaneous mark of defiance against the enforcement of rationality and predictability in the city. Stripped of pretension, the place of urban decay is the expression of the primordial substance of the built environment, a pre-rational geographic niche that has won over the authority.

The placelessness in the *terrain vague* of Rue Paradis and Boulevard Noir would constitute contemptuous space to mainstream community, while to a few like Sartre or Solà-Morales it expresses the authenticity of primordial, unformed urban environment. To the mainstream, no doubt, a disheveled place of abandon in a city is an urban twilight zone to avoid. To Sartre, however, contemptuous is the entirety of dominant city-form. In Sartre's literary rendition, the embodiment of overt automation across urban space, flaunted against nature whose spontaneity is shackled in a niche, such as a fenced and gated urban park, is a sinister undertone to the city's uniformity. It is the automated city, universal and uniform, that constitutes the loss of place:

> They come out of their offices after their day of work, they look at the houses and squares with satisfaction, they think it's their city, a good, solid bourgeois city. They aren't afraid, they feel at home. All they have ever seen is trained water running from taps, light which fills bulbs when you turn on the switch, half-breed, bastard trees held up with crutches. They have proof, a hundred times a day, that everything happens mechanically, that the world obeys fixed, unchangeable laws. In a vacuum all bodies fall at the same rate of speed, the public park is closed at 4 p.m. in winter, at 6 p.m. in summer, lead melts at 335 degrees centigrade, the last streetcar leaves Hotel de Ville at 11.05 p.m. They are peaceful, a little morose, they think about Tomorrow, that is to say, simply, a new today; cities have only one day at their disposal and every morning it comes back exactly the same. (Sartre 1938/1964: 158)

It is emblematic that Sartre touches on the different notions of placelessness and the loss of place, by means of his philosophical novel, *Nausea*, where these traits emerge through description of concrete spatial configurations, rather than being analyzed, defined or explained. Within the city, placelessness as the authentic and primordial urban substance has no clear rationale or explanation. Yet in contrast to the loss of place expressed by the blatancy of urban automation, uniformity and superscale of the mainstream, the placelessness of Rue Paradis and Boulevard Noir is not only a linguistic antipode to the mainstream, but also the rejection of, and the refusal to submit to the mechanistic rituals of the metropolis.

Benjamin, in the Marxist stage of his late twenties, had alluded to urban rituals in his enigmatic fragment of 1921, entitled, *Capitalism as a Religion*. While difficult to read as a coherent whole, Benjamin's fragment alludes to his observation of the German society, urbanized and strongly secular in cities, as a community guided by cultic premises of capital and money. Although, in hindsight of a century, it is easy to dismiss such an outlook, there are two unspoken, but leading concepts in Benjamin's fragment that are distinctly relevant precisely within the context of twentieth century's industrial and postindustrial city. These are the related notions of power and ritual, most pertinent to the state capitalism of Marxist governments.

The Communist Manifesto (1848/1964) of Karl Marx and Friedrich Engels was a blueprint, a masterplan of sorts, for the compulsory, mass displacement of human beings for the sake of

> combination of agriculture with manufacturing industries; gradual abolition of all the distinction between town and country by a more equable distribution of the populace over the country. (Marx and Engels 1848/1964: 94)

This blueprint was faithfully followed in the Soviet Union of the 1930s by forced collectivization of peasantry, in its monumental failure illustrating the futility of violence associated with the implementation of a fanciful, fictitious scheme. Translated into series of Soviet five-year plans, the first two in the series (1928–32, 1933–37) focused on heavy industry, expropriation of all private land and forcible agricultural co-operatives. The rigidity and the ruthless execution of the five-year plans led to famines in the Soviet Union, particularly in Ukraine, resulting in the deaths of millions of people.

In chapter 'Feuerbach. Opposition of the Materialist and Idealist Outlook' of their book, *The German Ideology*, Marx and Engels emphasized the communist premise of complete erasure of differences between town and country, urban and rural habitat and demeanor:

> The abolition of the antagonism between town and country is one of the first conditions of communal life, a condition which again depends on a mass of material premises and which cannot be fulfilled by the mere will, as anyone can see at the first glance. [...] The separation of town and country can also be understood as the separation of capital and landed property, as the beginning of the existence and development of capital independent of landed property—the beginning of property having its basis only in labour and exchange. (Marx and Engels 1846/1970: 69)

The implementation of the scheme to separate individuals and families from their land and property, *en route* to the messianic egalitarian society of communism, ended up in Soviet-style state capitalism of the 1950s and beyond. Prior to its implosion and ultimate demise in the late twentieth century, the Soviet Marxist state was run by an oligarchy of autocrats supported by armies of apparatchiks, police and informers. Suppressing traditional religion in favor of its own Marxist orthodoxy, the communist project had exiled millions into gulags near the Arctic Circle and forced the rest to participate in public rituals of marches under giant banners featuring Politburo chiefs, along with images of high priests of the religion, Marx and Engels. Medieval processions featuring religious banners of Catholic brotherhoods found their match in the atheistic creed of Marxism that punctuated seasons by rituals of forced citizens' marches. Commonplace were initiation rites for 10-year-olds joining the Young Pioneers organization at children ceremonies complete with military-style drills. The shipwreck of the Marxist delusionary masterplan for an entire society had left behind millions of lives destroyed physically or traumatized emotionally, not to speak of the material loss to individuals, families and the entire society.

On a much smaller scale in the west, twentieth-century urban masterplans have sealed the historical retreat from early urbanism of street vendors and market squares into a Trans-Atlantic urban failure, exemplified by Le Corbusier's schemes focused on speed and the residential tower, and by North America's suburban sprawl often exemplifying uniformity and monotonicity in geographic space. One of the power expressions of twentieth century's urban masterplans had been the conformity in automotive superscale of speed. "City made for speed is city made for success" was the heroic slogan of Le Corbusier, the great modern architect (Le Corbusier 1929/1996). Throughout history, access and encounter came hand in hand in the city where the street and the market square had facilitated both functions. Twentieth-century city-form, in its reliance on automation and speed, and the attendant repeating attempts to preserve spatial coherence, safety and security in its midst, had largely detached the attributes of urban access and human encounter from one another. Twentieth century's Myth of the Rational City, through automation and the superscale of height and speed, has been an attempt at a surprise-free city—automated, uniform and universal, at times also celebrating a master designer behind it.

High-speed urban commuting, the diurnal cycle of humans through metropolitan space in industrial and postindustrial age, is a prominent facet of the metropolis that startlingly fits Benjamin's observation of capitalism. In *Capitalism as Religion* Benjamin writes:

> Capitalism is the celebration of the cult *sans reve et sans merci* [without dream or mercy]. There are no "weekdays." There is no day that is not a feast day, in the terrible sense that all its sacred pomp is unfolded before us; each day commands the utter fealty of each worshipper. (Benjamin 1996: 288–290)

Urban commuting becomes a ritual through the invariance in its timeless repetition. It echoes Sartre's words, "Cities have only one day at their disposal and every morning it comes back exactly the same." Urban commuting implies "careful choreography [as well as] bodily discipline frequently performed in unison," the exact definition of a ritual as given by religions scholar, Catherine Bell (1997: 152–3). The urban commuters, universal across the western metropolis, have been described thus:

> They will seldom talk to each other [...]. When the train comes, their behavior is like some highly ritualized dance. It is plain that most of them know where they will be sitting and there is an etiquette and code of behavior which is fascinating to observe. They go through this same ritual every morning and it is clearly a satisfying and reassuring one to them, or they would not repeat it day after day. (Webb 2016: 68)

Commuting and urban passenger transportation industry, on the one hand, and suburban home construction and development industries, on the other hand, have created an economic feedback loop leading to twentieth-century sprawl of the suburbs in the western metropolis that continues to date. Housing and automobile industry constitute a critical mass of the North American consumer sector, compelling the government to guarantee their ongoing existence in the economy. This translates to enduring suburban sprawl and growth of the public and private passenger transport industries. Taken together these trends have led to the loss of place in the North American metropolis. Metropolitan transportation infrastructure, the universal manifestation of the feedback between residential suburbs and the urban economy, has set the stage for the ritual of metropolitan commuting, the passenger becoming a passive actor in the tragedy of the loss of place in the metropolis.

5.3 Eternal Return in the *Arcades Project*: The Coevolution of Archetypes

The ritual of urban commuting, unique to late modernity, deems the streetscape almost obsolete. The transformation of the metropolis into de Beauvoir's city without streets is the gradual, living experience of the loss of place. Historically, of course, urban streets and roads have only seldom been the sole subject of traffic and optimization of transits. When almost every one in the city attempts to optimize his or her own returns, the end result is not an optimized urban environment but a kind of urban superconscious (Akkerman 2016: 170). Yet only in the past hundred years have streets and roads come to be defined by optimization of transits for commuters, and ceased to be spatial medium of encounter. To Kundera, the transformation of streets and roads onto automotive conduits corresponds to the automatism of humans in the modern metropolis, both feeding into each other:

> Road: a strip of ground over which one walks. A highway differs from a road not only because it is solely intended for vehicles, but also because it is merely a line that connects one point with another. A highway has no meaning in itself; its meaning derives entirely from the two points that it connects.
>
> A road is a tribute to space. Every stretch of road has meaning in itself and invites you to stop. A highway is the triumphant devaluation of space, which thanks to it has been reduced to a mere obstacle to human movement and a waste of time.
>
> Before roads and paths disappeared from the landscape, they had disappeared from the human soul: man stopped wanting to walk, to walk on his own feet and to enjoy it. What's more, he no longer saw his own life as a road, but as a highway: a line that led from one point to another, from the rank of captain to the rank of general, from the role of wife to the role of widow. Time became a mere obstacle to life, an obstacle that had to be overcome by ever greater speed. (Kundera 1990: 223)

The ritual of motorized commuting has brought about the loss of place in streets of twentieth century's metropolis, helping to transform the urban square onto an intersection, and the entire city onto "a function of circulation, and of circuits" (Deleuze and Guattari 1980/2004: 477). The primordial space Solà-Morales and Sartre found in the alien, placeless urban niche of the *terrain vague*, as the other face of the city, is the opposite of the structured space of the planned city. Such contrariety seems to have a reflection in the predictability of the urban diurnal cycle of which the ritual of commuting is one facet, against the unstructured timing of sudden events and random occurrences throughout the city. The cyclical foundation of structured time of the mind *and* the metropolis is the vestigial archetype of the Eternal Return, much as *Axis mundi* is the archetypal reference of structured space.

In *Existence, Space and Architecture* (1971) Norberg-Schulz makes the case for *Axis mundi* as the initial, paradigmatic condition of place, as a space structured by the human being:

> The vertical, therefore, has always been considered the sacred dimension of space [...] The axis mundi is thus more than a center of the world, it represents a connection between the three cosmic realms [...] The simplest model of man's existential space is, therefore, a horizontal plane pierced by a vertical axis. (Norberg-Schulz 1971: 21)

Norberg-Schulz draws on the work of Mircea Eliade, *The Sacred and the Profane* (1961/1987). Eliade points out that *the Axis mundi is the underlying notion upon which place is conditioned inasmuch as it allows for a mythical Center to be articulated:*

> The three cosmic levels – earth, heaven, underworld – have been put in communication. [...T]his communication is sometimes expressed through the image of a universal pillar, *axis mundi*, which at once connects and supports heaven and earth and whose base is fixed in the world below. Such a cosmic pillar can be only at the very center of the universe, for the whole habitable world extends around it. Here, then, we have [...] the 'system of the world' prevalent in traditional societies: (a) a sacred place constitutes a break in the homogeneity of space; (b) this break is symbolized by an opening by which a passage from one cosmic region to another is made possible (from heaven to earth and vice versa; from earth to the underworld); (c) communication with heaven is expressed by one or another of certain images, all of which refer to the axis mundi: pillar (cf. the *universalis columna*), ladder (cf. Jacob's ladder), mountain, tree, vine, etc.; (d) around this cosmic axis lies the world (=our world), hence the axis is located "in the middle," at the "navel of the earth"; it is the Centre of the World. (Eliade 1961/1987: 36–37)

The spatial notion of a center is also the link of the *Axis mundi* with Eternal Return. The center and the axle presume not only vertical construction, but also horizontal surroundings, the inscription of a circle, and thus also cyclicality—the overt, structured expression of time. The Myth of the Eternal Return has in its common archaic transformation the round open space of the Neolithic solstice ritual, and later in the Bronze Age through to the Renaissance, the image of the *Ouroboros*, the circular symbol of a serpent eating its own tail. This is also how Norberg-Schulz perceives the initial condition of *place*: "A place is basically 'round'" (Norberg-Schulz 1971: 20).

In his *Arcades Project* Benjamin addresses the Myth of the Eternal Return almost always by dismissing it as a remnant of ancient superstition. Yet the Myth of the Eternal Return keeps haunting him, and he returns to it time and again, as if the myth was proving its own vitality to Benjamin. Ultimately, Benjamin (1933/1999: 118) brings in Nietzsche who had no qualm accepting the myth in its most orthodox version, in Aphorism 341 of his *Gay Science*:

> How, if some day or night a demon were to sneak after you into your loneliest loneliness and say to you, "This life as you now live it and have lived it, you will have to live once more and innumerable times more; and there will be nothing new in it, but every pain and every joy and every thought and sigh and everything immeasurably small or great in your life must return to you-all in the same succession and sequence– even this spider and this moon-light between the trees, and even this moment and I myself. The eternal hourglass of existence is turned over and over, and you with it, a dust grain of dust."
>
> Would you not throw yourself down and gnash your teeth and curse the demon who spoke thus? Or did you once experience a tremendous moment when you would have answered him, "You are a god, and never have I heard anything more godly." (Nietzsche 1882/974: 273–274)

The archetype of the Eternal Return has been formative in the cognitive background of religion as well as science. The Hindu religion, as one example, espouses cyclicality of the universe in the *Rigveda*, one of the four sacred *Vedas*, the canonical texts of Hinduism composed between 1500 and 1200 BCE. According to the *Rigveda* time and space are considered infinite and cyclical (Mittal and Thursby 2012: 284–285). Citing James Allen (1988: 21), Geraldine Pinch traces the Eternal Return to the Egyptian deity Atum that impregnates self, giving birth to the twins Shu and Tefnut:

> As the "father and mother" of the gods, Atum was the ultimate divine and royal ancestor. [...] He conceived and gave birth to the first two-gendered deities [who] sometimes appear to be the male and female aspects of Atum [...] As part of establishing the divine order, Shu and Tefnut also become two different types of time. "Shu is Eternal Recurrence and Tefnut is Eternal sameness." This began a great cycle in which everything had to change to survive and yet everything remained fundamentally the same. [...] At the beginning and end of each of the great cycles of existence, Atum took the form in the primeval waters as a snake or an eel. (Pinch pp. 64–111)

As an archetype, Eternal Return has been affirmatively constructive in the formulation of scientific hypotheses where cyclicality is a key concept. While supported by only a scant empirical evidence some of the cyclicality hypotheses are clearly formed through intuition inspired by the Myth of the Eternal Return. Precession of the equinoxes had been postulated by Hipparchus of Nicaea (c. 190–120 BCE), but his accurate statement regarding oscillation of the Earth rotation axis over a cycle of some 26,000 years was an extrapolation by way of acumen, as he was in possession of only some limited observational data going back to Mesopotamia, just two millennia earlier.

Precession of the equinoxes, changes in axial tilt of the Earth and small changes in the eccentricity of the elliptic orbit of the Earth around the sun, were postulated by Joseph Alphonse Adhémar and independently by James Croll in the nineteenth century, as having cumulative effect on cyclical variation of Earth's climate over millennia. Mathematical description of the joint effects of precession and changes in eccentricity and axial tilt upon millennial cycles of climate change were later formulated in a theory of climate cycles by the Serbian geophysicist and astronomer Milutin Milankovich (1879–1958). Adhémar, a French mathematician, in paving the way to Milankovich in his 1842 book *Revolutions of the Sea* had advanced the hypothesis that precession of the equinoxes affects the melting of polar ice caps and thereby leads to a cyclical catastrophic rise in sea levels. Human intuition aroused by the parable of Eternal Return, rather than empirical observation, played a dominant role in the forming of the Milankovich theory and its nineteenth-century precursors.

Similar intuitive drive regarding recurring aeonian cycles is behind the hypothesis by Michael Rampino and Ken Caldeira (2015) regarding periodic mass extinctions conjectured as linked to the earth's position in relation to our galaxy. The authors put forward a hypothesis of periodic impact and extinction events in intervals of 26 million years whereby they argue that 11 apparent impact events had occurred in the last 260 million years, including the one that eradicated the dinosaurs 66 million years ago.

The archetypal essence of the Eternal Return could probably be no better expressed than in the ancient verses of the Book of Koheleth, in the Old Testament:

> The words of Koheleth, the son of David, king in Jerusalem.
> Vanity of vanities, saith Koheleth; vanity of vanities, all is vanity.
> What profit hath man of all his labour wherein he laboureth under the sun?
> One generation passeth away, and another generation cometh; and the earth abideth for ever.
> The sun also ariseth, and the sun goeth down, and hasteth to his place where he ariseth.
> The wind goeth toward the south, and turneth about unto the north; it turneth about continually in its circuit, and the wind returneth again to its circuits.

> All the rivers run into the sea, yet the sea is not full; unto the place whither the rivers go, thither they go again.
> All things toil to weariness; man cannot utter it, the eye is not satisfied with seeing, nor the ear filled with hearing.
> That which hath been is that which shall be, and that which hath been done is that which shall be done; and there is nothing new under the sun.
> Is there a thing whereof it is said: 'See, this is new'? – it hath been already, in the ages which were before us. (Koheleth 1: 1–11)

Quite analogous to the placeless space—the *terrain vague*—Eternal Return too exudes air of meaninglessness. Both placelessness and meaninglessness, as the initial conditions of urban space and time, constitute an existential void, hollowness. Filling or moulding every single void in the urban space, forcing a meaning upon it through instrumental usefulness, does not necessarily lead to betterment of the urban space or the community occupying it. Rather, it points to an ingrained drive to ignore the past:

> There is no remembrance of them of former times; neither shall there be any remembrance of them of latter times that are to come, among those that shall come after. (Koheleth 1: 12)

5.4 LIA and the Scientific Revolution: Echoes of Iron Age Cold Epoch and the Axial Age?

Arguably, repetitiveness and universality are common tenets of ritual as well as of large segments of science. The two myths *Axis mundi* and Eternal Return, on this view, appear as evolutionary common ancestors to both religious ritual and to, at least some, science. If both religious ritual and some scientific thought constitute common descent from the two myths, it would seem only a small step to confer on the environ/*mental* progress through the archetypal pair, the Eternal Return and the *Axis mundi*, a flair of biotic process.

There is no clear linear link between this archetypal pair originating in the Paleolithic, its feminine feature representing cyclicality, movement and change, its masculine facet standing for constancy. Yet the coevolution of the two facets had presumably produced the environmental gender parables of the Garden and the Citadel later during the late Neolithic and the Bronze Age. The ensuing unfolding of gendered facets of city-form had been expressed in the open-air public agora and forum, on the one hand, and the fort and the walled city, on the other hand, during the Iron

Age and beyond. Further through medieval times and the Renaissance, against the emergence of market squares and accreted towns, the emergence of planned or ideal communities, and the articulation of the concept of the ideal city, begin to proliferate. And finally, during urban modernity, the feminine *terrain vague* as contemptable, decaying space, against attempts at a clockwork metropolis of a masterplan, define the dialectical framework for the contemporary metropolis.

In a propitious insight, Elisabeth de Bièvre had advanced her concept of the urban *subconscious* as the "sum of physical circumstances [...] and historical events, experienced collectively by a group of people living for several generations in the same environment" (de Bièvre 1995). De Bièvre concept is instrumental in discerning the historical transmutations of the archetypal pair of the Eternal Return and *Axis mundi*, as components of such urban subconscious. Accordingly, the built environment and humans within it, do not merely interact (Chap. 1), but—extending on Freud and Lefebvre (Sects. 3.2 and 5.1)—they embody a singular mind-city composite. The archetypal pair, Eternal Return and *Axis mundi*, as the initial gender projection upon the *natural* environment—the earth and the sky, has thus been transmuting throughout the symbiosis of humans and their minds with their built environments, across historic time to this very day.

The two environmental archetypes could be said to be in an ongoing process of coevolution, unceasingly molding the mind-city composite. On this view, *both* Descartes and Locke were right: although the mind of the newborn is not entirely a *tabula rasa*, empirical experience continually impacts paradigms ingrained in the mind and carries them forward through historic time. The two founders of the Enlightenment, mutual adversaries who never met each other, fathered modern rationalism and British empiricism not only through inspiration by their contemporaneous city-form (Sect. 3.4), but through their symbiotic immersion within it.

As pointed out already (Sect. 3.4), the perceptual contrast in the built environment of early modernity has very likely led to the virtually simultaneous discovery of coordinate geometry by Descartes, Girard Desargues and Pierre de Fermat. A 1922 study by sociologist and statistician William Ogburn and the demographer Dorothy Swaine Thomas pointed to almost 150 major inventions and discoveries that were made independently by two or more individuals at the same time (Ogburn and Thomas 1922). In 1961 Robert Merton had suggested further that "the pattern of independent multiple discoveries in science is in principle the dominant pattern, rather than a subsidiary one" (Merton 1961). There has been no alternative account of simultaneous multiple discoveries that would meaningfully vary from the explanation given by Ogburn and Thomas, namely, that simultaneous multiple discoveries were due to commonalities in "mental ability and cultural preparation."

A more pointed, environmental explanation can be sought in an example offered by Tetsurō Watsuji in reflecting upon 'cold' in his work, *Climate and Culture*, where he envisages a pedestrian stepping onto the street in cold weather. In her feeling cold the pedestrian makes no perceptual contact with any specific object, points out Watsuji: the pedestrian simply steps into a milieu (Watsuji 1988: 3–5). The feeling of cold characterizes the urban environment as a whole: the cold is shared among people within the same urban environment. Identical features of the urban environment, in its contrast between aligned and erratic streetscapes, were shared by Descartes, Desargues and de Fermat. It was the streetscape environment, specifically the grid pattern of streets in the *new towns*, on the backdrop of the Romanesque streetscape, that had led to the simultaneous discovery of coordinate geometry.

Identical patterns in climate *and* the built environment, shared among peoples, had constituted a critical epistemological backdrop to the formation of views and concepts that shaped philosophy and political ideology at the end of the LIA, one of its coldest periods, and the city-form emerging from Haussmann's masterplan for Paris. Modern city-form was able to offer excessive shielding from sunlight due to bad weather, offering relative comfort, but also leading to attendant prevalence of mood and mental disorder due to sunlight deprivation. During the nineteenth century, the close of the LIA, there had been a significant rise of lunatic asylums in Europe alluding thus also to a possible higher prevalence of mental disorder. Whereas the turn of the century saw only a few hundred individuals in asylums in England and France combined, "by the late 1890s and early 1900s, this number had risen to the hundreds of thousands" at which time also "German speaking countries housed more than 400 public and private sector asylums" (Shorter 1998: 34).

The near-simultaneous emergence of modern Existentialism and Marxism ought to be seen in this light. The two philosophical and ideological approaches are jointly a single modern example of psychosocial traits associated with identical urbanization patterns, enveloped in North European weather at the end of the LIA. Emerging from a continuing shielding from open air, along with increasingly automated and surprise-free city-form, has been the philosophy of despair and meaninglessness, and the ideology of rage and belligerence (Akkerman 2014). At the end of LIA, dejection and anguish had become features of Existentialism, while the advocacy of violent overthrow of ruling elites by the suffering working poor became the defining moment of Marxism.

The association between depression and violence, repeatedly brought up over the past half-century in psychiatric literature (Perlis et al. 2004), could hardly be ignored in the near-simultaneous emergence of both Existentialism and Marxism. It has been suggested for some time now that a physiological relation exists between depression and violence. Insufficient levels of serotonin, a hormone secreted in the brain with sun exposure, have been linked to depression as well as to violence (Krakowski 2003). Mental disorder in northern Europe at the end of the LIA, and in its aftermath, ought to be seen within this context: shielding from bad weather meant also insufficient exposure of Europeans to moderate ultraviolet-B solar radiation that induces production of vitamin D in skin and the release of endorphins, including serotonin (Juzeniene and Moan 2012). At the close of the LIA, the nineteenth century gave birth to European Existentialism, the philosophy of anguish, as well as to Marxism, an ideology of violence. Anguish as being behind the rise of European Existentialism, and violence at the arrival of Marxism, cannot be detached from prevalence of mood or mental disorder, that was also accompanied by rising strife in cities. Existentialism and Marxism, in their emergence at the same place and at the same time, seem to have a common source in the environmental context of weather and the built environment.

Analogous to such near-simultaneity had been cases of *identical* ideas emerging at *different* places at the same time. Axial Age civilizations serve as one such major specimen. In his book, *Origins and Goals of History*, Karl Jaspers observes common quest for meaning through the East and the Occident across the half-millennium, 800-200 BCE, that he had coined, Axial Age (Jaspers 1953; Schwartz 1975). At the doorstep of the Roman Empire and Christianity, basic tenets of later religious and philosophical thought of the West had emerged through the northern hemisphere in Axial Age civilizations, from China to Europe, south of the 50th parallel and largely along the Tropic of Cancer. Extending Jaspers' Axial geographic corridor through the Americas, the main urbanizing civilizations during the first millennium BCE, would include also the Maya and the Olmec.

Much of the Axial Age period overlaps the Iron Age Cold Epoch, sometimes referred to also as the Sub-Atlantic Cold Period, a cooling period of about 900–300 BCE (Behringer 2010: 57–59). Urbanization and advances in agriculture may have been linked to this cooling pattern. In China iron plowshares were invented in the course of the sixth century BCE (Greenberger 2006: 11–12), and Chinese irrigation systems were built along with associated hydraulic engineering networks during this period (Needham and Bray 1984: 138–161). In the Indus Valley the Axial Age coincides with new urban settlements arising at the Indo-Gangetic plain (Samuel 2008: 42–48). And in Ancient Greece Hippocrates of Kos (460–370 BCE) wrote the treatise, *On Airs, Waters and Places*, during the coldest part of this cooling epoch.

The coincidental emergence of similar patriarchal religions during the IACE throughout the Eurasian region is a prime example of the perceptual outcome in the observation of rotating circumpolar stars, the other result being the widespread use of the wheel among Eurasian peoples during IACE. Millennia prior to the Axial Age, the same coincidence between multifold inventions of the wheel, along with detection of rotating circumpolar stars, observable from the Great Plains of North America, from various high places and North Atlantic coastlines in Europe, led to prehistoric medicine wheels, and to cup-and-ring carvings near, or within, the same lines of geographic latitude. In other words, ardent observers of the nightly skies at locations spread along or near the same geographic latitude would notice identical sky patterns, the most impressive one being the nightly firmament rotating round the incumbent North Star (Chap. 2).

During the two-millennial period, 1500 BCE–500 CE Kochab (Beta Ursae Minoris) was the incumbent North Pole star. The geographical band between the Tropic of Cancer and the 45th parallel, approximately, is the earth's region from which, topography allowing, a sufficiently bright North Star, and adjoining circumpolar stars, are visible to casual onlooker. Kochab is observable today by the naked eye, and there is no reason to believe that this was not the case 3000 or 4000 years ago (Howse 1986; Haynie 2014). The revolving nightly firmament, and other sky patterns common to observers along roughly the 45th parallel through Eurasia, had occurred when wheels, axles, carts and whorls were common and familiar tools.

During the Iron Age Cold Period of 900–300 BCE, the use of agriculture and shelter construction would have intensified, carts and wagons becoming major technological advances in building construction. Within such a context assigning the nightly sky revolution round the North Star to a Grand Designer of the universe, was a plausible explanation.

It was this climatic, geographical and technological configuration that gave rise in China to the myth of the Jade Emperor, the Heavenly Grandfather, master of the universe. In the Indus Valley, Vishvakarman was the grand architect of the universe. Ahura mazda was the sky god of Zoroastrianism and of the Achaemenid Empire of Persia. And demiurge was the Platonic artisan fashioning and maintaining the physical universe. The geographic belt that hosted Venus figurines during the Upper Paleolithic, and the cup-and-ring marks during the Neolithic (Chap. 2), largely coincides also with the Eurasian corridor where similar cosmogonic myths had emerged during the IACE. Identical nightly sky patterns, accessibility in viewing them and existing technological tools had led to similar results of mythmaking in the common human drive for explanation (Akkerman 2015).

Two thousand years ago, in his treatise *De Opificio Mundi* (*On the Creation of the World*), Philo of Alexandria articulated a version of the Argument from Design for the existence of God, as an analogy between the ideal city of an architect's wisdom and the perfection of the universe:

> It is manifest also, that the archetypal seal, which we call that world which is perceptible only to the intellect, must itself be the archetypal model, the idea of ideas, the Logos of God, already occupied in the creation of the world; for neither is a city, while only perceptible to the intellect, anything else but the reason of the architect, who is already designing to build one perceptible to the external senses, on the model of that which is so only to the intellect. (*De Opificio Mundi* VI in Yonge 1854: 8)

Philo's Neo-Platonic city-universe analogy gives way to another, more bewildering parallel: the implicit notion of a similitude between the Grand Designer of the universe, and the master builder here on earth. Two millennia later, Henrik Ibsen's play, *The Master Builder*, shows the crisis of a human builder-designer realizing that his ambition toward eternity is forever challenged by others who are coming to replace him, with the same ambition. Ibsen's play alludes to an eternal loop where builder's ingrained desire for godliness and eternity through the construction of an edifice, a tower, becomes a tragic project akin to the Biblical Tower of Babel: the master builder is destroyed by the emergence of another one, doomed to follow in the same path.

5.5 The Roman Warm Period and Late Modernity Juxtaposed: An Epilogue

The Enlightenment and modernity had furnished the ascent of the rational mind, epitomized by science and critical thinking as the ultimate authority, displacing religion from the position of proverbial supremacy. Yet quite similar to the medieval status of Christianity emerging from late western antiquity, authority of the rational mind had become the absolute criterion of veracity and judgment. The Scientific Revolution and the Enlightenment, in raising the rational mind to the status of foremost authority, had displaced religion as a vestige of bygone age, much as Christianity replaced paganism during late antiquity and early Middle Ages in Europe. The genealogies of displacement of one universalist authority by another appear to have the commonality of inaugurating conduct toward worldly progress that is presented as the ousting of archaic obsolescence. Religions as well as the Scientific Revolution could be said to have been the agents of such intellectual and cultural displacement.

The rise of Christianity occurred between the second and fifth centuries, CE, during the second half of a climatic interval known as the Roman Warm Period, c. 250 BCE–400 CE, and the Scientific Revolution took place over the period, c. 1550–1700, toward the latter part of the LIA. The struggle between Christianity and secularism, as two cultural junctures of displacement, was at least to some extent reflected also in scuffles for civic power during the Middle Ages and early modernity. Within the territory of the former western Roman Empire and adjacent geographic regions, the millennium-long period between the rise of Christianity and the advent of the Scientific Revolution, was marked by competing sovereignties of power between and within cities or city-states, secular nobility, overlords and royalty, on the one hand, and the pope along with territorial authorities of the archbishops, bishops and clergy on the other hand.

Of concern ought to be similarities that begin to emerge between the second half of the Roman Warm Period and late modernity. Already at the end of the twentieth century there were some distinguished voices comparing the Early Middle Ages with late modernity. Hedley Bull had called the rising postmodernity 'neomedieval' due to parallels in civic and political arrangements in Europe (Bull 1977/2012: 245–6), and Habermas called the entire last segment of modernity 'refeudalization' due to what he saw as the appropriation of public sphere by private interests (Habermas 1962/1991: 141–158, 196–210).

It is too early to say whether patterns of demographic and climatic change in the North Atlantic region in the past 100 years constitute commencement of a period that will resemble centuries of late antiquity and early Middle Ages, spanning the Roman Warm Period and its immediate sequel. But if the last 100 years are any clue then both demographic and climatic traits of late modernity fall within the Bull-Habermas fold. The last 200 or 300 years of the Roman Warm Period along with the demographic sequel immediately following, and the hundred years of late modernity that came on the heels of the LIA, are unequal intervals but the demographic and climatic patterns of the two intervals, so far, resonate similarly.

Across the same geographic area of west-central Europe, the demographic similarity between the end of the Roman Warm Period and late modernity, stems from the comparison of population decline of the western Roman Empire during the lengthy period of its demise, against the dramatic decline of natural population growth in post WWII Europe after two decades of Baby Boom. The demographic similarity of the two periods is quite striking. Due to epidemics, violence from invasions and poor nutrition that surrounded the Sack of Rome of 410, and the ultimate collapse of the western Roman Empire half a century later, the European part of the Empire experienced population decline from about 23 million at the turn of the Common Era, to about 18.3 million in 350 CE (Russell 1972: 25–71). For comparison, in much of Europe after the 1950s' Baby Boom, birthrates began to decline, gradually to the extent that at the turn of the twenty-first century European populations, with the exception of south-eastern Europe, were below replacement levels.

In other words, natural growth (births minus deaths) in much of Europe at the turn of the new millennium has been negative. The only reason European government statistics show population increase is due to some migration from south-eastern Europe, but mainly due to migration from Asia and Africa (Frejka and Sobotka 2008; Favell 2008). The medieval Migration Period began toward the end of the Roman Warm Period, c. 300 CE, with violent movements of Germanic tribes from the north of Europe into the territory of the Western Empire, the Sack of Rome being the prominent mark of the early stage of the Migration Period. The last half a century of demographic change in Europe alludes to a parallel with about two centuries of demographic change between the decline of the Roman Empire and the start of the medieval Migration Period.

The Germanic migration movement into the Western Empire, in initial stages of the Migration Period, c. 300–500 CE, came at the end of the Roman Warm Period and half-a-millennium after the prosperous Axial Age of the IACE. The southern and eastern migration pressure upon central-western Europe during the early twenty-first century to the present time has come also on the heels of prosperous western Europe from the second half of the twentieth century to present. The entire medieval Migration Period, c. 300–700 CE, was marked by violent movement to west-central Europe by tribes from eastern and northern Europe, North Africa and Asia, whereas current migration pressure into roughly the same area stems from Africa and Asia, and to a lesser extent, from Eastern and south-eastern Europe. Early IACE till the turn of the Common Era was a period of population increase and urbanization throughout Greece and the Roman Republic (Bairoch 1988: 79–92) while current migration pressure comes on the heels of modern demographic transition in Europe that had occurred between the eighteenth and nineteenth centuries, marked also by rapid increase of urban populations (Bairoch 1988: 238–240).

The Roman Warm Period had followed the IACE, and in comparison, climate following the LIA has been, so far, also in a general trend of rising average temperatures. Climatic comparison of the Roman Warm Period with late modernity, therefore, is instructive not only in pointing to similarities between the two periods, but also due to the anomaly of average temperature rise during the twentieth century up until the present time. A widely recognized study of summer temperatures reconstruction over the last two millennia shows that the Roman Warm Period was warmer than the period from the mid-twentieth century through to the present time, but it shows also a general trend of cooling between late antiquity and the turn of the twentieth century. The study by Jan Esper et al. (2012) indicates that following the hot summers of the Roman Warm Period, a summer cooling trend in Europe persisted till the turn of twentieth century, then precipitously rising. Average summer temperatures, from the turn of the twentieth century up until now, are still below those of the Roman Warm Period, but significantly contravene the long-term trend of cooling over the last two millennia (Fig. 5.1).

Esper's temperature reconstruction was based on X-ray measurements of changes in cell-wall density of living and subfossil pine trees from 587 locations at 14 lakes and 3 lakeshore sites in northern Scandinavia. The reconstruction of summer temperatures revealed "a long-term cooling

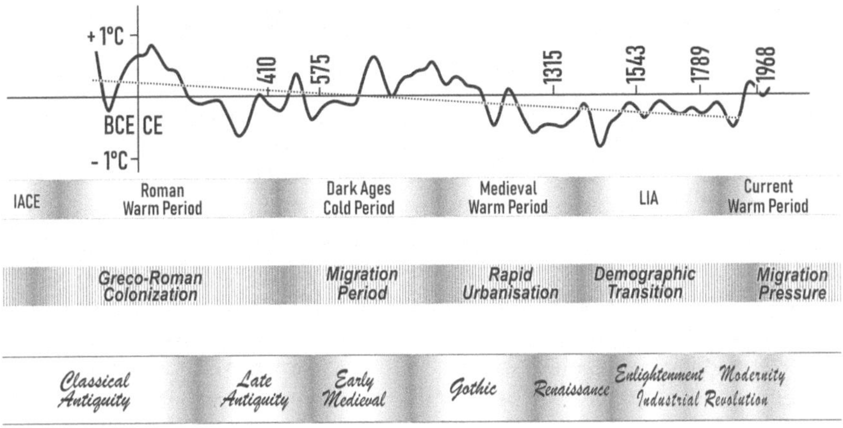

Fig. 5.1 Climate, demographic and cultural history periods, against estimated average summer temperatures, Scandinavia, 250 BCE–2000 CE. (Source: Esper et al. 2012. Drawn by Paul Van Pul)

Years on time axis:

410 – Sack of Rome by the Germanic Visigoths
575 – Visigoths invade Spain in aftermath of the Plague of Justinian (541–542 CE)
1315 – Great Famine starts, lasts till 1317, first of major Late Medieval European crises
1543 – Copernicus, *De revolutionibus orbium coelestium*, onset of the Scientific Revolution
1789 – Storming of the Bastille, French Revolution starts, lasts till 1799
1968 – Garrett Hardin publishes *Tragedy of the Commons*, on unsustainable development
Note: Average summer temperature, 1950–2000, was taken by Esper et al. (2012) as the standard (0°C) against which historical temperatures were measured

trend of –0.31°C per 1,000 years (±0:03°C) over the 138 B.C.–A.D. 1900 period" (Esper et al. 2012). The cooling of –0.3°C per millennium was attributed by Esper's team to gradual changes in the position of the sun "resulting from long-term oscillations of orbital configurations (Milankovitch 1941), as an important driver of Holocene climate." Though the researchers did not claim their reconstruction was definitive, they argued that the reconstruction reflects long-term changes in "orbital configurations" that have continually reduced average summer temperatures through the northern hemisphere over the past two millennia. The generally warming trend of the past hundred years could thus not be due to solar variability, but more likely, the result of rising greenhouse gas concentrations in the atmosphere.

The end of the Western Empire marked the onset of Feudalism in Europe, accompanied by overall cultural, economic and social decline on the Continent. Along with the Bull-Habermas civic parallel drawn between the two historic periods, a readily posed question could be asked, whether the early Middle Ages and the ascent of Christianity, as sequels to Axial Age and IACE, are not a telltale omen to postmodernity, itself a sequel to the LIA, the Enlightenment and the victory of the rational mind epitomized in the enterprise of technological innovation.

Within the urban setting, the early promise of modernity in the masterplans of the early twentieth century had been for a well-functioning city. At the end of modernity a considerable feature of the urban environment is the malfunctioning city. From the midst of euphoria of modernity, an increasingly disturbing pattern has emerged, whereby the postmodern 'technopolis' has become

> the milieu in which the tie between information and human purpose has been severed, [...] information appears indiscriminately, directed at no one in particular, in enormous volume at high speeds, and disconnected from theory, meaning, or purpose. (Postman 1993: 67–70)

It may be difficult to assert with any certitude the extent to which environmental conditions, other than those directly ensuing from technology, have played a role in such a disturbing pattern. But it would seem that the pattern of postmodernity, as described in the excerpt by Neil Postman, largely corresponds to the loss of place in the metropolis. Human scale is no match to superscale of late modernity, and human imagination is inconsistent with the uniformity of suburban sprawl. Meanwhile, superscale, uniformity and automation have largely done away with memories of place in the geographic space of the metropolis, save fossilized museum exhibits and perhaps some low-income neighborhoods in the inner city, where developer interests have not yet reached.

In his essay, *The Work of Art in the Age of Mechanical Reproduction*, Benjamin (1935/2008) suggests that much of what is left to human comfort in twentieth century's western civilization are automated settings, those that in the metropolis are reinforcing human automatism but hardly a source of happiness or contentment. Automation, uniformity and superscale, hailed in Thomas More's *Utopia* or in Descartes' *Discourse on the*

Method, were during the sixteenth and seventeenth centuries the striking marks of modernity, but most importantly, they provided a perceptual contrast to the built environment of the Middle Ages. Half a millennium later, no such contrasting opportunity exists in the contemporary metropolis. Urban dysfunction due to these three attributes of the postmodern metropolis carries the overt consequences of human conflict in accidents, traffic congestion, wanton crime or lineups. A subtle, yet long-lasting impact is urban pollution. The effect of urban air pollutants on greenhouse gases and global warming have been known for some time now (e.g., Ramanathan and Feng 2009), but recently also evidence of adverse impact of urban air pollution on cognitive ability has emerged (Zhang et al. 2018).

It would be a fallacy to say that, in comparison, urban light pollution is less acute than air pollution. The myth of *Axis mundi*, for the first time in human history, has become a vague and vanishing paradigm. The perceptual link between the earth and the sky exists no longer in the metropolis. Light pollution has deemed the view of a starry night sky an exceptional event occurring during a blackout or at special forays far outside the city, rather than an ongoing perceptual impact spurring human imagination and fantasy. The end of the nineteenth century may have been the last segment of modernity where starry sky and the city were bound together. The breaking of the link between the sky and the built environment may have come to signal a revolution, not in the sense of dramatic displacement of a competing regime, but the drifting return to an archaic incumbency of the *Ouroboros* archetypal cycle (Fig. 5.2).

In his essay, Benjamin addresses painting as a vestige of human authenticity. It is unclear whether Benjamin had any specific painters in mind, but Vincent van Gogh certainly fills the bill. Mental disorder of van Gogh made him cut off much of his left ear in December 1888 and brought him to a lunatic asylum in the town of Saint-Rémy-de-Provence in southern France, near the city of Arles. Perhaps it was during one of his hallucinatory periods during his stay in the asylum, that van Gogh created in the early second half of June 1889 his painting, *The Starry Night*. Albert Boime researched the painting and gave the following account of the night sky of *The Starry Night*:

> *The Starry Night* of that period was painted by gaslight and clearly shows the Big Dipper. Van Gogh wrote to his friend Boch that he had painted *The Starry Night* with the Great Bear, the northern constellation containing the seven stars that form the Dipper. A comparison of the painting with planetarium

Fig. 5.2 *Ouroboros*, by Lucas Jennis (1590–1630)

> reconstruction of the night sky over Arles at approximately 9 p.m. indicates that van Gogh tilted the constellation upward, but in his letter to Boch he seems to have adjusted the angle and also included Polaris. (Boime 1984)

Van Gogh thus distorted the nightly view of the sky by forcing the North Star into the painting. This alone may not have been noteworthy if it was not for another deliberate distortion by the artist, one that gives his painting a momentous dimension. Writes Boime:

> [...W]e must [account] for two objects van Gogh could not have seen – the curious serpentine movement in the center of the night sky and the human-made church of Saint-Martin which lay to the northwest of the asylum and could not have been viewed through the window. (Boime 2008: 20)

The painting connects the Neolithic with modernity: the sky and the city. The *Axis mundi* is formed by the North Star, the diminished spire of the church, as well as a cypress tree in the left-hand side of the painting:

> The cypress tree alone of all the terrestrial elements breaks through confinement and shoots into the sky like a projectile to touch the stars. Because the cypress tree is an evergreen, it was planted around tombs in Mediterranean countries as a symbol of immortality [...T]he cypress tree is Vincent's own secularized version of the Eiffel Tower that enables him to make contact with the stars. (Boime 2008: 42)

The serpentine movement in the center of the night sky could be, as Boime suggests, the whirlpool nebula, the first spiral galaxy, discovered in the eighteenth century. It is, however, as likely that van Gogh alludes to *Ouroboros*, the serpent of the Eternal Return. The images of stars in the painting, as concentric circles, almost like cup-and-ring marks, reinforce such interpretation. The archetypal bond, or its late loss, between the built form and the sky is either celebrated or mourned in this painting. Though van Gogh's intent may forever remain unresolved, his imagery alludes to a closure of an epoch. In *The Starry Night* van Gogh's tortured mind took a glimpse of the sky with the human-built environment and linked the Neolithic with the closing of modernity. Van Gogh may have done this as an ominous yearning, on behalf of us all.

Bibliography

Akkerman, Abraham. 2014. Winter-Cities and Mood Disorder: Observations from European City-Form at the End of Little Ice Age. *Trames Journal of the Humanities and Social Sciences* 18 (1): 19–37.

———. 2015. Myths of the North and Origins of City-Form: Some Reflections Across History and Prehistory. *Journal of Architecture and Urbanism* 39 (3): 165–175.

———. 2016. *Phenomenology of the Winter-City: Myth in the Rise and Decline of Built Environments*. New York/Dordrecht/London: Springer.

Allen, James B. 1988. Genesis in Egypt. In *The Philosophy of Ancient Egyptian Creation Accounts*. New Haven: Yale University Press.

Bairoch, Paul. 1988. *Cities and Economic Development: From the Dawn of History to the Present*. Chicago: University of Chicago Press.

Baudelaire, Charles. 1857/2010. *The Flowers of Evil & Paris Spleen: Selected Poems*. Mineola: Dover.

Behringer, Wolfgang. 2010. *A Cultural History of Climate*. Cambridge: Polity Press.

Bell, Catherine. 1997. *Ritual: Perspectives and Dimensions*. New York: Oxford University Press.

Benjamin, Walter. 1933/1999. *The Arcades Project*. Cambridge, MA: Harvard University Press.

———. 1935/2008. *The Work of Art in the Age of Mechanical Reproduction*. London: Penguin.

———. 1996. Capitalism as Religion. In *Walter Benjamin: Selected Writings*, ed. Howard Eiland and Gary Smith, vol. 1, 1913–1926. Cambridge, MA: Harvard Belknap.

Boime, Albert. 1984. Van Gogh's *Starry Night*: A History of Matter and a Matter of History. *Arts Magazine* 59 (4): 86–103.

———. 2008. *Revelation of Modernism: Responses to Cultural Crises in Fin-de-siècle Painting*. Columbia/London: University of Missouri Press.

Bull, Hedley N. 1977/2012. *The Anarchical Society: A Study of Order in World Politics*. New York: Palgrave Macmillan.

Chambers, Ross. 1993. *The Writing of Melancholy: Modes of Opposition in Early French Modernism*. Chicago: University of Chicago.

de Bièvre, Elisabeth. 1995. The Urban Subconscious: The Art of Delft and Leiden. *Art History* 18 (2): 222–252.

de Solà-Morales, Ignasi. 1995/2014. Terrain Vague. In *Terrain Vague: Interstices at the Edge of the Pale*, ed. Patrick Barron and Manuela Mariani, 24–30. New York: Routledge.

Deleuze, Gilles, and Pierre-Félix Guattari. 1980/2004. *Capitalism and Schizophrenia—Volume II: A Thousand Plateaus*. London/New York: Continuum.

Eliade, Mircea. 1954. *Cosmos and History: The Myth of the Eternal Return*. New York: Harper & Brothers.

———. 1961/1987. *The Sacred and the Profane: The Nature of Religion*. New York: Harcourt.

Esper, Jan, David C. Frank, Mauri Timonen, Eduardo Zorita, Rob Wilson, Jürg Luterbacher, Steffen Holzkämper, Nils Fischer, Sebastian Wagner, Daniel Nievergelt, Anne Verstege, and Ulf Büntgen. 2012. Orbital Forcing of Tree-Ring Data. *Nature Climate Change* 2 (12): 862–866.

Favell, Adrian. 2008. The New Face of East–West Migration in Europe. *Journal of Ethnic and Migration Studies* 34 (5): 701–716.

Foucault, Michel. 1967/1986. Of Other Spaces. Trans. Jay Miskowiec. *Diacritics* 16 (1): 22–27.

Frejka, Tomas, and Tomas Sobotka. 2008. Fertility in Europe: Diverse, Delayed and Below Replacement. In *Childbearing Trends and Policies in Europe, Book 1*, ed. Tomas Frejka, Jan Home, and Tomas Sobotka, 15–46. Norderstedt: Max-Planck-Gesellschaft.

Frynta, Emmanuel. 1960. *Kafka and Prague*. London: Batchworth Press.

Gasarian, Gerard. 1997. 'Le Cygne' of Baudelaire. In *Understanding Les Fleurs Du Mal: Critical Readings*, ed. William J. Thompson, 122–132. Nashville: Vanderbilt University Press.

Greenberger, Robert. 2006. *The Technology of Ancient China*. New York: Rosen Publishing Group.

Habermas, Jürgen. 1962/1991. *The Structural Transformation of the Public Sphere: An Inquiry into a Category of Bourgeois Society*. Cambridge: MIT Press.

Haynie, Donald T. 2014. Nabta Playa in the Development of Science and Technology. *Technology & Innovation* 16 (1): 63–73.

Hildebrandt, S. 2008. Capital Punishment and Anatomy: History and Ethics of an Ongoing Association. *Clinical Anatomy* 21 (1): 5–14.

Howse, Derek. 1986. Navigation and Astronomy: The First Three Thousand Years. *Renaissance and Modern Studies* 30 (1): 60–86.

Jaspers, Karl. 1953. *The Origin and Goal of History*. Trans. Michael Bullock. London: Routledge and Keegan Paul.

Juzeniene, Asta, and Johan Moan. 2012. Beneficial Effects of UV Radiation Other Than Via Vitamin D Production. *Dermatoendocrinology* 4 (2): 109–117.

Krakowski, Menahem. 2003. Violence and Serotonin: Influence of Impulse Control, Affect Regulation, and Social Functioning. *The Journal of Neuropsychiatry and Clinical Neurosciences* 15: 294–305.

Kundera, Milan. 1990. *Immortality*. New York: Grove Weidenfeld.

Le Corbusier. 1929/1996. A Contemporary City. In *The City Reader*, ed. Richard T. LeGates and Frederic Stout, 336–344. London/New York: Routledge.

Lefebvre, Henri. 1974/1991. *The Production of Space*. Oxford: Blackwell.

———. 1996. The Right to the City. In *Writings on Cities*, ed. Eleonore Kofman and Elizabeth Lebas, 147–160. Oxford: Blackwell.

———. 2003. *The Urban Revolution*. Minneapolis/London: University of Minnesota Press.

Marx, Karl, and Friedrich Engels. 1848/1964. *The Communist Manifesto*. New York: Simon & Schuster.

———. 1970. *The German Ideology*. New York: International Publishers.

Matthew, A.W. 2012. Assessing the Impacts of Phenotypic Plasticity on Evolution. *Integrative and Comparative Biology* 52 (1): 5–15.

Merton, Robert K. 1961. The Role of Genius in Scientific Advance. *New Scientist* 12 (259): 306–308.

Milankovitch, Milutin. 1941. *Kanon der Erdbestrahlung und seine Anwendung auf das Eiszeitenproblem*. Belgrad: Royal Serbian Academy.

Mittal, Sushil, and Gene Thursby. 2012. *Hindu World*. New York: Routledge.

Needham, Joseph, and Francesca Bray. 1984. *Science and Civilisation in China: Volume 6, Biology and Biological Technology*. Cambridge/New York: Cambridge University Press.

Nietzsche, Friedrich. 1882/1974. *The Gay Science*. New York: Vintage.

Norberg-Schulz, Christian. 1971. *Existence, Space & Architecture*. New York: Praeger Publishers.

———. 1980. *Genius loci: Towards a Phenomenology of Architecture*. New York: Rizzoli.

Ogburn, William F., and Dorothy Thomas. 1922. Are Inventions Inevitable? A Note on Social Evolution. *Political Science Quarterly* 37 (1): 83–98.

Perlis, Roy H., J.W. Smoller, M. Fava, J.F. Rosenbaum, A.A. Nierenberg, and G.S. Sachs. 2004. The Prevalence and Clinical Correlates of Anger Attacks During Depressive Episodes in Bipolar Disorder. *Journal of Affective Disorders* 79 (1): 291–295.

Philo of Alexandria c. 30/1854. On the Creation of the World (*De opificio mundi*). In *The Works of Philo Judaeus*, ed. and trans. Charles D. Yonge, 1–52. London: Henry G. Bohn.

Pinch, Geraldine. 2002. *Egyptian Mythology: A Guide to the Gods, Goddesses, and Traditions of Ancient Egypt*. Oxford: Oxford University Press.

Postman, Neil. 1993. *Technopoly: The Surrender of Culture to Technology*. New York: Knopf.

Ramanathan, V., and Y. Feng. 2009. Air Pollution, Greenhouse Gases and Climate Change: Global and Regional Perspectives. *Atmospheric Environment* 43 (1): 37–50.

Rampino, Michael R., and Ken Caldeira. 2015. Periodic Impact Cratering and Extinction Events Over the Last 260 Million Years. *Monthly Notices of the Royal Astronomical Society* 454 (4): 3480–3484.

Relph, Edward. 1976. *Place and Placelessness*. London: Pion.

Rousseau, Jean-Jacques. 1782/1861. *The Confessions of Jean-Jacques Rousseau*. London: Reeves and Turner.

Russell, Josiah Cox. 1972. Population in Europe. In *The Fontana Economic History of Europe*, The Middle Ages, ed. Carlo M. Cipolla, vol. 1. Glasgow: Collins/Fontana.

Samuel, Geoffrey. 2008. *The Origins of Yoga and Tantra*. Cambridge: Cambridge University Press.

Sartre, Jean-Paul. 1938/1964. *Nausea*. Trans. Lloyd Alexander. New York: New Directions.

Schwartz, Benjamin I. 1975. The Age of Transcendence, *Daedalus* (Special Issue: *Wisdom, Revelation and Doubt: Perspectives on the First Millennium BCE*) 104 (2): 1–7.

Shinar, David. 2017. *Traffic Safety and Human Behavior*. Bingley: Emerald Publishing.

Shorter, Edward. 1998. *A History of Psychiatry: From the Era of the Asylum to the Age of Prozac*. New York: Wiley.

Stanek, Lukasz. 2011. *Henri Lefebvre on Space: Architecture, Urban Research, and the Production of Theory*. Minneapolis/London: University of Minnesota Press.

Watsuji, Tetsurō. 1988. *Climate and Culture: A Philosophical Study*. New York: Greenwood Press.

Webb, Simon. 2016. *Commuters: The History of a British Way of Life*. Barnsley: Pen & Sword.

Zhang, Xin, Xi Chen, and Xiaobo Zhang. 2018. The Impact of Exposure to Air Pollution on Cognitive Performance. *Proceedings of the National Academy of Sciences of the United States of America* 115 (37): 9193–9197.

Index

A. Akkerman, *Philosophical Urbanism*,
https://doi.org/10.1007/978-3-030-29085-6

FSC
www.fsc.org
MIX
Papier | Fördert gute Waldnutzung
FSC® C083411

Kolibri 360 GmbH
Ferdinand-Jühlke-Straße 7
99095 Erfurt, Deutschland
produktsicherheit@kolibri360.de